LONDON MATHEMATICAL SOCIETY STUDENT TEXTS

Managing Editor: Professor D. Benson,
Department of Mathematics, University of Aberdeen, UK

London Mathematical Society Student Texts 84

Random Graphs, Geometry and Asymptotic Structure

MICHAEL KRIVELEVICH
Tel-Aviv University

KONSTANTINOS PANAGIOTOU
University of Munich

MATHEW PENROSE
University of Bath

COLIN MCDIARMID
University of Oxford

Edited by

NIKOLAOS FOUNTOULAKIS
University of Birmingham

DAN HEFETZ
University of Birmingham

CAMBRIDGE
UNIVERSITY PRESS

Shaftesbury Road, Cambridge CB2 8EA, United Kingdom

One Liberty Plaza, 20th Floor, New York, NY 10006, USA

477 Williamstown Road, Port Melbourne, VIC 3207, Australia

314–321, 3rd Floor, Plot 3, Splendor Forum, Jasola District Centre, New Delhi – 110025, India

103 Penang Road, #05–06/07, Visioncrest Commercial, Singapore 238467

Cambridge University Press is part of Cambridge University Press & Assessment,
a department of the University of Cambridge.

We share the University's mission to contribute to society through the pursuit of
education, learning and research at the highest international levels of excellence.

www.cambridge.org
Information on this title: www.cambridge.org/9781107136571

© Cambridge University Press & Assessment 2016

First published 2016

A catalogue record for this publication is available from the British Library

Library of Congress Cataloging-in-Publication data
Names: Krivelevich, Michael. | Fountoulakis, Nikolaos, editor. | Hefetz, Dan, editor.
Title: Random graphs, geometry, and asymptotic structure / Michael Krivelevich,
Tel-Aviv University [and three others]; edited by Nikolaos Fountoulakis, University of
Birmingham, Dan Hefetz, University of Birmingham.
Description: Cambridge : Cambridge University Press, 2016. |
Series: London Mathematical Society student texts; 84 |
Includes bibliographical references and index.
Identifiers: LCCN 2015043108| ISBN 9781107136571 (hardback) |
ISBN 9781316501917 (pbk.)
Subjects: LCSH: Random graphs. | Geometry, Algebraic. | Topology.
Classification: LCC QA166.17.R37 2016 | DDC 511/.5–dc23
LC record available at http://lccn.loc.gov/2015043108

ISBN   978-1-107-13657-1   Hardback
ISBN   978-1-316-50191-7   Paperback

Contents

Editors' introduction

The theory of random graphs was established during the 1950s through the pioneering work of Gilbert and subsequently of Erdős and Rényi who set its foundations. Since then, the theory has been developed vastly and is by now a central area of combinatorics. Numerous, often unexpected, ramifications have emerged, which link it to diverse areas of mathematics such as number theory, combinatorial optimization and probability theory. Since its beginning, the study of geometric and topological aspects of random graphs has become the meeting point between combinatorics and areas of probability theory, such as percolation theory and stochastic processes. Nowadays, this interface has been consolidated through numerous deep results. This has led to applications in other scientific disciplines including telecommunications, astronomy, statistical physics, biology and computer science, as well as much more recent developments such as the study of social and biological networks.

The present book is the outcome of a short course that took place at the School of Mathematics of the University of Birmingham in August 2013 and was supported by the London Mathematical Society and the Engineering and Physical Sciences Research Council. Its aim was to provide a concise overview of recent trends in the theory of random graphs, ranging from classical structural problems to geometric and topological aspects, and to introduce the participants to new powerful complex–analytic techniques and stochastic models that have led to recent breakthroughs in the field.

The theory of random graphs is nowadays part and parcel of the education of any young researcher entering the fascinating world of combinatorics. However, due to their interdisciplinary nature, the geometric and structural aspects of the theory often remain an obscure part of the education of young researchers. Moreover, the theory itself, even in its most basic forms, is often considered quite advanced to be part of undergraduate curricula, and those interested, usually learn it mostly through self-study, covering a lot of its fundamentals but not much of the more recent developments. The present book provides a self-contained and concise introduction to recent developments and techniques for classical problems in the theory of random graphs. Moreover, it covers geometric and topological aspects of the theory of random graphs and introduces the reader to the diversity and depth of the methods that have been invented in this context. Emphasis is given to powerful complex–analytic approaches as well as to sophisticated tools based on stochastic processes. To the best of our knowledge, it is the first time that such diverse approaches are put together in a single volume.

This book consists of four chapters that complement each other. These chapters are aimed to be mostly self-contained so as to be particularly accessible to researchers with limited background in the theory of random graphs. They have been developed by Prof. M. Krivelevich, Prof. K. Panagiotou, Prof. M. D. Penrose and Prof. C. McDiarmid, respectively. Each of these four academics is an internationally acclaimed expert in the field covered in their chapter.

The first chapter covers some recent developments regarding the classical problem of finding long paths and cycles in random graphs. Paths and cycles have always been between the most central and well-studied notions in graph theory. It is thus only natural they have become one of the foci of attention of the theory of random graphs, resulting in many beautiful results and inspiring methods whose importance reaches far beyond probabilistic combinatorics. This chapter reviews several basic results and approaches in long cycles and Hamiltonicity in random graphs, covering both classical theorems and recent developments.

The second chapter covers properties of random objects from graph classes with structural constraints, such as planar graphs. The corresponding random graph models are usually much harder to work with than the classical binomial random graph model. The standard methodology that is applied in this context is the use of generating functions and analytic techniques. In this chapter, a new approach based on random sampling techniques that is both technically significantly simpler and more widely applicable is presented. The chapter starts with a general presentation of the method with an application to random trees. It then proceeds with the rigorous development of the theory of combinatorial species so that general classes can be decomposed and sampled. This part contains a basic development of the so-called Boltzmann sampling, as it has been developed by Flajolet and co-authors. This method is applied to determine the typical block structure of graphs from certain classes as well as their degree sequence.

The third chapter focuses on the theory of finite graphs with nodes placed randomly in a Euclidean space and edges added to connect those pairs of nodes that are close to one another. As an alternative to classical random graph models, these geometric graphs are relevant to the modeling of real-world networks having spatial content, arising in applications in areas such as wireless communications, classification, epidemiology and astronomy. The following topics are covered in this chapter: connectivity and Hamiltonicity, distributional limit theorems, the maximal and minimal degree.

The fourth and final chapter provides an overview of the theory of random graphs from restricted classes via a *combinatorial* point of view. In particular, random graphs from minor-closed classes and bridge-addable classes are also discussed.

We would like to thank Dr. Allan Lo and Dr. Elisabetta Candellero for assisting us with the smooth running of the short course. Finally, we would like to thank the London Mathematical Society and the Engineering and Physical Sciences Research Council for their generous financial support and their invaluable assistance with various organizational matters.

1

Long paths and Hamiltonicity in random graphs

Michael Krivelevich

1 Introduction

Long paths and Hamiltonicity are certainly among the most central and researched topics of modern graph theory. It is thus only natural to expect that they will take a place of honor in the theory of random graphs. And indeed, the typical appearance of long paths and of Hamilton cycle is one of the most thoroughly studied directions in random graphs, with a great many diverse and beautiful results obtained over the past fifty or so years.

In this survey we aim to cover some of the most basic theorems about long paths and Hamilton cycles in the classical models of random graphs, such as the binomial random graph or the random graph process. By no means should this text be viewed as a comprehensive coverage of results of this type in various models of random graphs; the reader looking for breadth should rather consult research papers or a recent monograph on random graphs by Frieze and Karoński [1]. Instead, we focus on simplicity, aiming to provide accessible proofs of several classical results on the subject and showcasing the tools successfully applied recently to derive new and fairly simple proofs, such as applications of the Depth First Search (DFS) algorithm for finding long paths in random graphs and the notion of boosters.

Although this chapter is fairly self-contained mathematically, basic familiarity and hands-on experience with random graphs would certainly be of help for the prospective reader. The standard random graph theory monographs of Bollobás [2] and of Janson et al. [3] certainly provide (much more than) the desired background.

This chapter is based on a mini-course with the same name, delivered by the author at the LMS-EPSRC Summer School on Random Graphs, Geometry, and Asymptotic Structure, organized by Dan Hefetz and Nikolaos Fountoulakis at the University of Birmingham in the summer of 2013. The author would like to thank the course organizers for inviting him to deliver the mini-course, and for encouraging him to create lecture notes for the course, which eventually served as a basis for the present chapter.

2 Tools

In this section we gather notions and tools to be applied later in the proofs. They include standard graph theoretic notation, asymptotic estimates of binomial coefficients, Chebyshev's and Chernoff's inequalities, and basic notation from random graphs (Section 2.1), as well as algorithmic and combinatorial tools – the DFS algorithm for graph exploration (Section 2.2), the so-called rotation-extension technique of Pósa, and the notion of boosters (Section 2.3).

2.1 Preliminaries

2.1.1 Notation and terminology

Our graph theoretic notation and terminology are fairly standard. In particular, for a graph $G = (V, E)$ and disjoint vertex subsets $U, W \subset V$, we denote by $N_G(U)$ the external neighborhood of U in G as: $N_G(U) = \{v \in V - U : v \text{ has a neighbor in } U\}$. The number of edges of G spanned by U is denoted by $e_G(U)$ and the number of edges of G between U and W is $e_G(U, W)$. When the graph G is clear from the context, we may omit G in the subscript in the above notation.

Path and cycle lengths are measured in edges.

When dealing with graphs on n vertices, we will customarily use N to denote the number of pairs of vertices in such graphs: $N = \binom{n}{2}$.

2.1.2 Asymptotic estimates

We will use the following, quite standard and easily proven, estimates of binomial coefficients. Let $1 \leq x \leq k \leq n$ integers. Then

$$\left(\frac{n}{k}\right)^k \leq \binom{n}{k} \leq \left(\frac{en}{k}\right)^k, \tag{2.1}$$

$$\frac{\binom{n-x}{k-x}}{\binom{n}{k}} \leq \left(\frac{k}{n}\right)^x, \tag{2.2}$$

$$\frac{\binom{n-x}{k}}{\binom{n}{k}} \leq e^{-\frac{kx}{n}}. \tag{2.3}$$

2.1.3 Chebyshev and Chernoff

Chebyshev's inequality helps to show concentration of a random variable X around its expectation, based on the first two moments of X. It reads as follows: let X be a random variable with expectation μ and variance σ^2. Then for any $a > 0$,

$$Pr[|X - \mu| \geq a\sigma] \leq \frac{1}{a^2}.$$

The following are very standard bounds on the lower and the upper tails of the binomial distribution due to Chernoff: If $X \sim Bin(n, p)$, then

- $Pr(X < (1-a)np) < \exp\left(-\frac{a^2 np}{2}\right)$ for every $a > 0$.
- $Pr(X > (1+a)np) < \exp\left(-\frac{a^2 np}{3}\right)$ for every $0 < a < 1$.

Another trivial, yet useful, bound is as follows: let $X \sim Bin(n, p)$ and $k \in \mathbb{N}$. Then

$$Pr(X \geq k) \leq \left(\frac{enp}{k}\right)^k.$$

Indeed, $Pr(X \geq k) \leq \binom{n}{k} p^k \leq \left(\frac{enp}{k}\right)^k$.

2.1.4 Random graph, asymptotic notation

As usual, $G(n, p)$ denotes the probability space of graphs with vertex set $\{1, \ldots, n\} = [n]$, where every pair of distinct elements of $[n]$ is an edge of $G \sim G(n, p)$ with probability p, independently of other pairs. For $0 \leq m \leq N$, $G(n, m)$ denotes the probability space of all graphs with vertex set $[n]$ and exactly m edges, where all such graphs are equiprobable: $Pr[G] = \binom{N}{m}^{-1}$. One can expect that the probability spaces $G(n, p)$ and $G(n, m)$ have many similar features; when the corresponding parameters are appropriately tuned: $m = Np$, accurate quantitative statements are available, see [2] and [3]. This similarity frequently allows us to prove a desired property for one of the probability spaces, and then to transfer it to the other one.

We will also address briefly the model $D(n, p)$ of directed random graphs, defined as follows: the vertex set is $[n]$, and each of the $n(n-1) = 2N$ ordered pairs $(i, j), 1 \leq i \neq j \leq n$ is a directed edge of $D(n, p)$ with probability p, independently from other pairs.

We say that an event $\mathcal{E}_n$ occurs with high probability, or **whp** for brevity, in the probability space $G(n, p)$ if $\lim_{n \to \infty} Pr[G \sim G(n, p) \in \mathcal{E}_n] = 1$. (Formally, one should rather talk about a sequence of events $\{\mathcal{E}_n\}_n$ and a sequence of probability spaces $\{G(n, p)\}_n$.) This notion is defined in other (sequences of) probability spaces in a similar way.

Let $k \geq 2$ be an integer, and assume that $0 \leq p, p_1, \ldots, p_k \leq 1$ satisfy $(1-p) = \prod_{i=1}^{k}(1 - p_i)$. Then the random graphs $G \sim G(n, p)$ and $G' = \bigcup_{i=1}^{k} G(n, p_i)$ have the exact same distribution. Indeed, it is obvious that each pair of vertices

$1 \leq i < j \leq n$ is an edge in both graphs G, G' independently of other pairs. In G, this edge does not appear with probability $1 - p$, and in order for it to not appear in G', it should not appear in any of the random graphs $G(n, p_i)$ – which happens with probability $\prod_{i=1}^{k}(1 - p_i) = 1 - p$, the same probability as in G. This very useful trick is called *multiple exposure* as it allows us to generate (to expose) a random graph $G \sim G(n, p)$ in stages, by generating the graphs $G(n, p_i)$ sequentially and then by taking their union. When the last probability p_k is much smaller than the rest, it is also called *sprinkling* – a typical scenario in this case is to expose the bulk of the random graph $G \sim G(n, p)$ first by generating the graphs $G(n, p_i)$, $i = 1, \ldots, k - 1$, to come close to a target graph property P, and then to add few random edges from the last random graph $G(n, p_k)$ (to sprinkle these few edges) to finish off the job.

2.2 Depth First Search and its applications for finding long paths

The *Depth First Search* is a well-known graph exploration algorithm, usually applied to discover connected components of an input graph. As it turns out, this algorithm is particularly suitable for finding long paths in graphs, and using it in the context of random graphs can really make wonders. We will see some of them later in the chapter.

Recall that the DFS (Depth First Search) is a graph search algorithm that visits all vertices of a (directed or undirected) graph. The algorithm receives a graph $G = (V, E)$ as an input; it is also assumed that an order π on the vertices of G is given, and the algorithm prioritizes vertices according to π. The algorithm maintains three sets of vertices, letting S be the set of vertices whose exploration is complete, T be the set of unvisited vertices, and $U = V \setminus (S \cup T)$, where the vertices of U are kept in a stack (the last-in, first-out data structure). It initializes with $S = U = \emptyset$ and $T = V$ and runs until $U \cup T = \emptyset$. At each round of the algorithm, if the set U is nonempty, the algorithm queries T for neighbors of the last vertex v that has been added to U, scanning T according to π. If v has a neighbor u in T, the algorithm deletes u from T and inserts it into U. If v does not have a neighbor in T, then v is taken out of U and is moved to S. If U is empty, the algorithm chooses the first vertex of T according to π, deletes it from T, and pushes it into U. In order to complete the exploration of the graph, whenever the sets U and T have both become empty (at this stage the connected component structure of G has already been revealed), we make the algorithm query all remaining pairs of vertices in $S = V$, not queried before. Figure 1.1 provides an illustration of applying the DFS algorithm.

Observe that the DFS algorithm starts revealing a connected component C of G at the moment the first vertex of C gets into (empty beforehand) U and completes discovering all of C when U becomes empty again. We call a

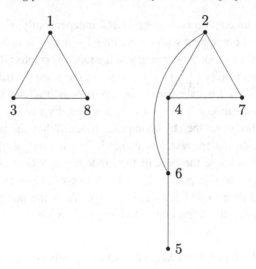

Step	S	U	T
0	$\emptyset$	$\emptyset$	$\{1,\ldots,8\}$
1	$\emptyset$	1	$\{2,\ldots,8\}$
2	$\emptyset$	$1,3$	$\{2,4,\ldots,8\}$
3	$\emptyset$	$1,3,8$	$\{2,4,\ldots,7\}$
4	$\{8\}$	$1,3$	$\{2,4,\ldots,7\}$
5	$\{3,8\}$	1	$\{2,4,\ldots,7\}$
6	$\{1,3,8\}$	$\emptyset$	$\{2,4,\ldots,7\}$
7	$\{1,3,8\}$	2	$\{4,5,6,7\}$
8	$\{1,3,8\}$	$2,4$	$\{5,6,7\}$
9	$\{1,3,8\}$	$2,4,6$	$\{5,7\}$
10	$\{1,3,8\}$	$2,4,6,5$	$\{7\}$
11	$\{1,3,5,8\}$	$2,4,6$	$\{7\}$
12	$\{1,3,5,6,8\}$	$2,4$	$\{7\}$
13	$\{1,3,5,6,8\}$	$2,4,7$	$\emptyset$
14	$\{1,3,5,6,7,8\}$	$2,4$	$\emptyset$
15	$\{1,3,4,5,6,7,8\}$	2	$\emptyset$
16	$\{1,\ldots,8\}$	$\emptyset$	$\emptyset$

Figure 1.1. Graph G with vertices labeled is on the left, and the protocol of applying the DFS algorithm to G is on the right. Observe that at any point of the algorithm execution, the set U spans a path in G.

period of time between two consecutive emptyings of U an *epoch*, each epoch corresponding to one connected component of G. During the execution of the DFS algorithm as depicted in Figure 1.1, there are two components in the graph G, and there are two epochs – the first is Steps 1–6 and the second is Steps 7–16.

The following properties of the DFS algorithm are immediate to verify:

(D1) at each round of the algorithm one vertex moves, either from T to U or from U to S;

(D2) at any stage of the algorithm, it has been revealed already that the graph G has no edges between the current set S and the current set T;

(D3) the set U always spans a path (indeed, when a vertex u is added to U, it happens because u is a neighbor of the last vertex v in U; thus, u augments the path spanned by U, of which v is the last vertex).

We now exploit the features of the DFS algorithm to derive the existence of long paths in expanding graphs.

Proposition 2.1 *Let k, l be positive integers. Assume that $G = (V, E)$ is a graph on more than k vertices, in which every vertex subset S of size $|S| = k$ satisfies $|N_G(S)| \geq l$. Then G contains a path of length l.*

Proof Run the DFS algorithm on G, with π being an arbitrary ordering of V. Look at the moment during the algorithm execution when the size of the set S of already processed vertices becomes exactly equal to k (there is such a moment as the vertices of G move into S one by one, until eventually all of them land there). By Property **(D2)** above, the current set S has no neighbors in the current set T, and thus $N(S) \subseteq U$, implying $|U| \geq l$. The last move of the algorithm was to shift a vertex from U to S, so before this move the set U was one vertex larger. The set U always spans a path in G, by Property **(D3)**. Hence G contains a path of length l. $\qquad\qquad\square$

Proposition 2.2 *[4] Let $k < n$ be positive integers. Assume that $G = (V, E)$ is a graph on n vertices, containing an edge between any two disjoint subsets $S, T \subset V$ of size $|S| = |T| = k$. Then G contains a path of length $n - 2k + 1$ and a cycle of length at least $n - 4k + 4$.*

Proof Run the DFS algorithm on G, with π being an arbitrary ordering of V. Consider the moment during the algorithm execution when $|S| = |T|$ – there is such a moment, by Property **(D1)**. Since G has no edges between the current S and the current T by Property **(D2)**, it follows by the proposition's assumption that both sets S and T are of size at most $k - 1$. This leaves us with the set U whose size satisfies: $|U| \geq n - 2k + 2$. Since U always spans a path by **(D3)**, we obtain a path P of desired length. To argue about a cycle, take the first and the last k vertices of P. By the proposition's assumption there is an edge between these two sets, this edge obviously closes a cycle with P, whose length is at least $n - 4k + 4$, as required. □

As we have hinted already, the DFS algorithm is well suited to handle directed graphs too. Similar results to those stated above can be obtained for the directed case. Here is an analog of Proposition 2.2 for directed graphs; the proof is the same, mutatis mutandis.

Proposition 2.3 *[4] Let $k < n$ be positive integers. Let $G = (V, E)$ be a directed graph on n vertices, such that for any ordered pair of disjoint subsets $S, T \subset V$ of size $|S| = |T| = k$, G has a directed edge from S to T. Then G contains a directed path of length $n - 2k + 1$ and a directed cycle of length at least $n - 4k + 4$.*

2.3 Pósa's Lemma and boosters

In this section we present yet another technique for showing the existence of long paths in graphs. This technique, introduced by Pósa in 1976 [5] in his research on Hamiltonicity of random graphs, is applicable not only for arguing about long paths, but also for various Hamiltonicity questions. And indeed, we will see its application in this context later.

In quite informal terms, Pósa's Lemma guarantees that expanding graphs not only have long paths, but also provide a very convenient backbone for augmenting a graph to a Hamiltonian one by adding new (random) edges. The fact that expanders are good for getting long paths is already not new to us – Propositions 2.1 and 2.2 are about just this. Pósa's Lemma, however, quantifies things somewhat differently, and as a result yields further benefits.

We start by defining formally the notion of an expander.

Definition 2.4 *For a positive integer k and a positive real α, a graph $G = (V, E)$ is a (k, α)-expander if $|N_G(U)| \geq \alpha |U|$ for every subset $U \subset V$ of at most k vertices.*

By the way of example, Proposition 2.1 can now be rephrased (in a somewhat weaker form – we now require the expansion of all sets of size *up to k*) as follows: if G is a (k, α)-expander, then G has a path of length at least αk. For technical reasons, Pósa's Lemma uses the particular case $\alpha = 2$.

The idea behind Pósa's approach is fairly simple and natural – one can start with a (long) path P, and then, using extra edges, perform a sequence of simple deformations (rotations) until it will be possible to close the deformed path to a cycle, or to extend it by appending a vertex outside $V(P)$; then this can be repeated if necessary to create an even longer path, or to close it to a Hamilton cycle. The approach is thus called naturally the *rotation-extension* technique.

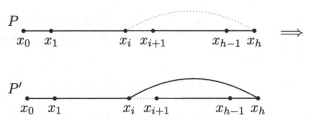

Figure 1.2. Elementary rotation is applied to a path P to get a new path P' with the same vertex set.

Formally, let $P = x_0 x_1 \ldots x_h$ be a path in a graph $G = (V, E)$, starting at a vertex x_0. Suppose G contains an edge (x_i, x_h) for some $0 \le i < h - 1$. Then a new path P' can be obtained by rotating the path P at x_i, that is, by adding the edge (x_i, x_h) and erasing (x_i, x_{i+1}). This operation is called an *elementary rotation* and is depicted in Figure 1.2. Note that the obtained path P' has the same length h and starts at x_0. We can therefore apply an elementary rotation to the newly obtained path P', resulting in a path P'' of length h, and so on. If after a number of rotations an endpoint x of the obtained path Q is connected by an edge to a vertex y outside Q, then Q can be extended by adding the edge (x, y).

The power of the rotation-extension technique of Pósa hinges on the following lemma.

Lemma 2.5 *[5] Let G be a graph, P be a longest path in G and $\mathcal{P}$ be the set of all paths obtainable from P by a sequence of elementary rotations. Denote by R the set of ends of paths in $\mathcal{P}$, and by R^- and R^+ the sets of vertices immediately preceding and following the vertices of R along P, respectively. Then $N_G(R) \subset R^- \cup R^+$.*

Proof Let $x \in R$ and $y \in V(G) \setminus (R \cup R^- \cup R^+)$, and consider a path $Q \in \mathcal{P}$ ending at x. If $y \in V(G) \setminus V(P)$, then $(x, y) \notin E(G)$, as otherwise the path Q can be extended by adding y, thus contradicting our assumption that P is a longest path. Suppose now that $y \in V(P) \setminus (R \cup R^- \cup R^+)$. Then y has the same neighbors in every path in $\mathcal{P}$, because an elementary rotation that removed one of its neighbors along P would, at the same time, put either this neighbor or y itself in R (in the former case $y \in R^- \cup R^+$). Then if x and y are adjacent, an elementary rotation applied to Q produces a path in $\mathcal{P}$ whose endpoint is a neighbor of y along P, while belonging itself to R, a contradiction. Therefore in both cases x and y are non-adjacent.   □

The following immediate consequence of Lemma 2.5 is frequently applied in Hamiltonicity problems in random graphs.

Corollary 2.6 *Let h, k be positive integers. Let $G = (V, E)$ be a graph such that its longest path has length h, but it contains no cycle of length $h + 1$.*

Suppose furthermore that G is a $(k,2)$-expander. Then there are at least $\frac{(k+1)^2}{2}$ non-edges in G such that if any of them is turned into an edge, then the new graph contains an $(h+1)$-cycle.

Proof Let $P = x_0 x_1 \ldots x_h$ be a longest path in G and let R, R^-, R^+ be as in Lemma 2.5. Notice that $|R^-| \leq |R|$ and $|R^+| \leq |R| - 1$, since $x_h \in R$ has no following vertex on P and thus does not contribute an element to R^+.

According to Lemma 2.5,

$$|N_G(R)| \leq |R^- \cup R^+| \leq 2|R| - 1 ,$$

and it follows that $|R| > k$. (Here the choice $\alpha = 2$ in the definition of a (k,α)-expander plays a crucial role.) Moreover, (x_0, v) is not an edge for any $v \in R$ (there is no $(h+1)$-cycle in the graph), whereas adding any edge (x_0, v) for $v \in R$ creates an $(h+1)$-cycle.

Fix a subset $\{y_1, \ldots, y_{k+1}\} \subset R$. For every $y_i \in R$, there is a path P_i ending at y_i, which can be obtained from P by a sequence of elementary rotations. Now fix y_i as the starting point of P_i and let Y_i be the set of endpoints of all paths obtained from P_i by a sequence of elementary rotations. As before, $|Y_i| \geq k+1$, no edge joins y_i to Y_i, and adding any such edge creates a cycle of length $h+1$. Altogether we have found $(k+1)^2$ pairs (y_i, x_{ij}) for $x_{ij} \in Y_i$. As every non-edge is counted at most twice, the conclusion of the lemma follows.   □

The reason we are after a cycle of length $h+1$ in the above argument is that if $h+1 = n$, then a Hamilton cycle is created. Otherwise, if the graph is connected, then there will be an edge e connecting a newly created cycle C of length $h+1$ with a vertex outside C. Then opening C up and appending e in an obvious way creates a longer path in G.

In order to utilize quantitatively the above argument, we introduce the notion of boosters.

Definition 2.7 *Given a graph G, a non-edge $e = (u,v)$ of G is called a* booster *if adding e to G creates a graph G', which is Hamiltonian or whose longest path is longer than that of G.*

Note that technically every non-edge of a Hamiltonian graph G is a booster by definition.

Boosters advance a graph towards Hamiltonicity when added; adding sequentially n boosters clearly brings any graph on n vertices to Hamiltonicity.

We thus conclude from the previous discussion:

Corollary 2.8 *Let G be a connected non-Hamiltonian $(k,2)$-expander. Then G has at least $\frac{(k+1)^2}{2}$ boosters.*

3 Long paths in random graphs

In this section we treat the appearance of long paths and cycles in sparse random graphs. We will work with the probability space $G(n,p)$ of binomial random graphs; analogous results for the sister model $G(n,m)$ can be either proven using very similar arguments or derived using available equivalence statements between the two models.

Our goal here is two-fold: we first prove that already in the super-critical regime $p = \frac{1+\epsilon}{n}$, the random graph $G(n,p)$ typically contains a path of length linear in n; then we prove that in the regime $p = \frac{C}{n}$, the random graph $G(n,p)$ typically has a path covering the proportion of vertices tending to 1 as the constant C increases. We will invoke the approaches and results developed in Section 2.2 (the DFS algorithm and its consequences) to achieve both of these goals, in fact in a rather short and elegant way.

3.1 Linearly long paths in the supercritical regime

In their groundbreaking paper [6] from 1960, Paul Erdős and Alfréd Rényi made the following fundamental discovery: the random graph $G(n,p)$ undergoes a remarkable phase transition around the edge probability $p(n) = \frac{1}{n}$. For any constant $\epsilon > 0$, if $p = \frac{1-\epsilon}{n}$, then $G(n,p)$ has **whp** all connected components of size at most logarithmic in n, while for $p = \frac{1+\epsilon}{n}$ **whp** a connected component of linear size, usually called the giant component, emerges in $G(n,p)$ (they also showed that **whp** there is a unique linear-sized component). The Erdős–Rényi paper, which launched the modern theory of random graphs, has had enormous influence on the development of the field. We will be able to derive both parts of this result very soon.

Although for the super-critical case $p = \frac{1+\epsilon}{n}$ the result of Erdős and Rényi shows a typical existence of a linear-sized connected component, it does not imply that a longest path in such a random graph is **whp** linearly long. This was established some 20 years later by Ajtai et al. [7]. In this section we present a fairly simple proof of their result. We will not attempt to achieve the best possible absolute constants, aiming rather for simplicity. Our treatment follows closely that of [8].

The most fundamental idea of the proof is to run the DFS algorithm on a random graph $G \sim G(n,p)$, constructing the graph "on the fly," as the algorithm progresses. We first fix the order π on $V(G) = [n]$ to be the identity permutation. When the DFS algorithm is fed with a sequence of i.i.d. Bernoulli(p) random variables $\bar{X} = (X_i)_{i=1}^N$, so that it gets its i-th query answered positively if $X_i = 1$ and answered negatively otherwise, the so-obtained graph is clearly distributed according to $G(n,p)$. Thus, studying the component structure of G can be reduced to studying the properties of the random sequence $\bar{X}$. This is a very useful trick, as it allows us to "flatten" the random graph by replacing an inherently two-dimensional structure (a graph) by a one-dimensional one (a sequence of bits). In particular (here and later we use the DFS algorithm-related notation of Section 2.2), observe crucially that as long as $T \neq \emptyset$, every positive answer to a query results in a vertex being moved from T to U, and thus after t queries and assuming $T \neq \emptyset$ still, we have $|S \cup U| \geq \sum_{i=1}^{t} X_i$. (The last inequality is strict in fact as the first vertex of each connected component is moved from T to U "for free," that is, without the need to get a positive answer to a query.) On the other hand, since the addition of every vertex, but the first one in a connected component, to U is caused by a positive answer to a query, we have at time t: $|U| \leq 1 + \sum_{i=1}^{t} X_i$.

The probabilistic part of our argument is provided by the following quite simple lemma.

Lemma 3.1 *Let $\epsilon > 0$ be a small enough constant. Consider the sequence $\bar{X} = (X_i)_{i=1}^{N}$ of i.i.d. Bernoulli random variables with parameter p.*

1. *Let $p = \frac{1-\epsilon}{n}$. Let $k = \frac{7}{\epsilon^2} \ln n$. Then **whp** there is no interval of length kn in $[N]$, in which at least k of the random variables X_i take value 1.*
2. *Let $p = \frac{1+\epsilon}{n}$. Let $N_0 = \frac{\epsilon n^2}{2}$. Then **whp** $\left| \sum_{i=1}^{N_0} X_i - \frac{\epsilon(1+\epsilon)n}{2} \right| \leq n^{2/3}$.*

Proof (1) For a given interval I of length kn in $[N]$, the sum $\sum_{i \in I} X_i$ is distributed binomially with parameters kn and p. Applying the Chernoff bound to the upper tail of $B(kn, p)$, and then the union bound, we see that the probability of the existence of an interval violating the assertion of the lemma is at most

$$(N - k + 1)Pr[B(kn, p) \geq k] < n^2 \cdot e^{-\frac{\epsilon^2}{3}(1-\epsilon)k} < n^2 \cdot e^{-\frac{\epsilon^2(1-\epsilon)}{3}\frac{7}{\epsilon^2}\ln n} = o(1),$$

for small enough $\epsilon > 0$.
(2) The sum $\sum_{i=1}^{N_0} X_i$ is distributed binomially with parameters N_0 and p. Hence, its expectation is $N_0 p = \frac{\epsilon n^2 p}{2} = \frac{\epsilon(1+\epsilon)n}{2}$, and its standard deviation is of order $\sqrt{n}$. Applying the Chebyshev inequality, we get the required estimate. □

Now we are ready to formulate and prove the main result of this section.

Theorem 3.2 *Let $\epsilon > 0$ be a small enough constant. Let $G \sim G(n, p)$.*

1. *Let $p = \frac{1-\epsilon}{n}$. Then **whp** all connected components of G are of size at most $\frac{7}{\epsilon^2} \ln n$.*
2. *Let $p = \frac{1+\epsilon}{n}$. Then **whp** G contains a path on at least $\frac{\epsilon^2 n}{5}$ vertices.*

In both cases, we run the DFS algorithm on $G \sim G(n, p)$ and assume that the sequence $\bar{X} = (X_i)_{i=1}^{N}$ of random variables, defining the random graph

$G \sim G(n,p)$ and guiding the DFS algorithm, satisfies the corresponding part of Lemma 3.1.

Proof (1) Assume to the contrary that G contains a connected component C with more than $k = \frac{7}{\epsilon^2} \ln n$ vertices. Let us look at the epoch of the DFS when C was created. Consider the moment inside this epoch when the algorithm has found the $(k+1)$-st vertex of C and is about to move it to U. Denote $\Delta S = S \cap C$ at that moment. Then $|\Delta S \cup U| = k$, and thus the algorithm got exactly k positive answers to its queries to random variables X_i during the epoch, with each positive answer being responsible for revealing a new vertex of C, after the first vertex of C was put into U in the beginning of the epoch. At that moment during the epoch only pairs of vertices touching $\Delta S \cup U$ have been queried, and the number of such pairs is therefore at most $\binom{k}{2} + k(n-k) < kn$. It thus follows that the sequence $\bar{X}$ contains an interval of length at most kn with at least k 1's inside – a contradiction to Property 1 of Lemma 3.1.

(2) Assume that the sequence $\bar{X}$ satisfies Property 2 of Lemma 3.1. We claim that after the first $N_0 = \frac{\epsilon n^2}{2}$ queries of the DFS algorithm, the set U contains at least $\frac{\epsilon^2 n}{5}$ vertices (with the contents of U forming a path of desired length at that moment). Observe first that $|S| < \frac{n}{3}$ at time N_0. Indeed, if $|S| \geq \frac{n}{3}$, then let us look at a moment $t \leq N_0$ where $|S| = \frac{n}{3}$ (such a moment surely exists as vertices flow to S one by one). At that moment $|U| \leq 1 + \sum_{i=1}^{t} X_i < \frac{n}{3}$ by Property 2 of Lemma 3.1. Then $|T| = n - |S| - |U| \geq \frac{n}{3}$, and the algorithm has examined all $|S| \cdot |T| \geq \frac{n^2}{9} > N_0$ pairs between S and T (and found them to be non-edges) – a contradiction. Let us return to time N_0. If $|S| < \frac{n}{3}$ and $|U| < \frac{\epsilon^2 n}{5}$ then, we have $T \neq \emptyset$. This means in particular that the algorithm is still revealing the connected components of G, and each positive answer it got resulted in moving a vertex from T to U (some of these vertices may have already migrated further from U to S). By Property 2 of Lemma 3.1 the number of positive answers at that point is at least $\frac{\epsilon(1+\epsilon)n}{2} - n^{2/3}$. Hence, we have $|S \cup U| \geq \frac{\epsilon(1+\epsilon)n}{2} - n^{2/3}$. If $|U| \leq \frac{\epsilon^2 n}{5}$, then $|S| \geq \frac{\epsilon n}{2} + \frac{3\epsilon^2 n}{10} - n^{2/3}$. All $|S||T| \geq |S| \left(n - |S| - \frac{\epsilon^2 n}{5}\right)$ pairs between S and T have been probed by the algorithm (and answered in the negative). We thus get:

$$\frac{\epsilon n^2}{2} = N_0 \geq |S| \left(n - |S| - \frac{\epsilon^2 n}{5}\right) \geq \left(\frac{\epsilon n}{2} + \frac{3\epsilon^2 n}{10} - n^{2/3}\right)\left(n - \frac{\epsilon n}{2} - \frac{\epsilon^2 n}{2} + n^{2/3}\right)$$

$$= \frac{\epsilon n^2}{2} + \frac{\epsilon^2 n^2}{20} - O(\epsilon^3)n^2 > \frac{\epsilon n^2}{2}$$

(we used the assumption $|S| < \frac{n}{3}$ in the second inequality above), and this is obviously a contradiction, completing the proof. $\square$

Let us discuss briefly the obtained result and its proof. First, given the probable existence of a long path in $G(n,p)$, that of a long cycle is just one short step further. Indeed, we can use sprinkling as follows. Let $p = \frac{1+\epsilon}{n}$ for small $\epsilon > 0$. Write $1 - p = (1 - p_1)(1 - p_2)$ with $p_2 = \frac{\epsilon}{2n}$; thus, most of the probability p goes into $p_1 \geq \frac{1+\epsilon/2}{n}$. Let now $G \sim G(n, p)$, $G_1 \sim G(n, p_1)$, $G_2 \sim G(n, p_2)$, we can represent $G = G_1 \cup G_2$. By Theorem 3.2, G_1 **whp** contains a linearly long path P. Now, the edges of G_2 can be used to close most of P into a cycle – there is **whp** an edge of G_2 between the first and the last $n^{2/3}$ (say) vertices of P.

The dependencies on ϵ in both parts of Theorem 3.2 are of the correct order of magnitude – for $p = \frac{1-\epsilon}{n}$ a largest connected component of $G(n, p)$ is known to be **whp** of size $\Theta(\epsilon^{-2}) \log n$ while for $p = \frac{1+\epsilon}{n}$ a longest cycle of $G(n, p)$ is **whp** of length $\Theta(\epsilon^2)n$ (see for example Chapter 6 of [2]).

Observe that using a Chernoff-type bound for the tales of the binomial random variable instead of the Chebyshev inequality would allow us to claim in the second part of Lemma 3.1 that the sum $\sum_{i=1}^{N_0} X_i$ is close to $\frac{\epsilon(1+\epsilon)n}{2}$ with probability exponentially close to 1. This would show in turn, employing the argument of Theorem 3.2, that $G(n, p)$ with $p = \frac{1+\epsilon}{n}$ contains a path of length linear in n with exponentially high probability, namely, with probability $1 - \exp\{-c(\epsilon)n\}$.

As we mentioned in Section 2.2, the DFS algorithm is applicable equally well to directed graphs. Hence essentially the same argument as above, with obvious minor changes, can be applied to the model $D(n, p)$ of random digraphs. It then yields the following theorem:

Theorem 3.3 *Let* $p = \frac{1+\epsilon}{n}$, *for* $\epsilon > 0$ *constant. Then the random digraph* $D(n, p)$ *has* **whp** *a directed path and a directed cycle of length* $\Theta(\epsilon^2)n$.

This recovers the result of Karp [9].

3.2 Nearly spanning paths

Consider now the regime $p = \frac{C}{n}$, where C is a (large) constant. Our goal is to prove that **whp** in $G(n, p)$, the length of a longest path approaches n as C tends to infinity. This too is a classical result due to Ajtai et al. [7], and independently due to Fernandez de la Vega [10]. It is fairly amusing to see how easily it can be derived using the DFS-based tools we developed in Section 2.2.

Theorem 3.4 *For every* $\epsilon > 0$ *there exists* $C = C(\epsilon) > 0$ *such that the following is true. Let* $G \sim G(n, p)$, *where* $p = \frac{C}{n}$. *Then* **whp** *G contains a path of length at least* $(1 - \epsilon)n$.

Proof Clearly we can assume $\epsilon > 0$ to be small enough. Let $k = \lfloor \frac{\epsilon n}{2} \rfloor$. By Proposition 2.2 it suffices to show that $G \sim G(n, p)$ contains **whp** an edge between every pair of disjoint subsets of size k of $V(G)$. For a given pair of disjoint sets S, T of size $|S| = |T| = k$, the probability that G contains no edges between S and T is exactly $(1 - p)^{k^2}$ (all k^2 pairs (s, t), $s \in S$

and $t \in T$, come out non-edges in G). Using the union bound, we obtain that the probability of the existence of a pair violating this requirement is at most

$$\binom{n}{k}\binom{n-k}{k}(1-p)^{k^2} < \binom{n}{k}^2 (1-p)^{k^2} < \left(\frac{en}{k}\right)^{2k} e^{-pk^2} < \left[\left(\frac{en}{k}\right)^2 \cdot e^{-\frac{Ck}{n}}\right]^k.$$

Recalling the value of k and taking $C = \frac{5\ln(1/\epsilon)}{\epsilon}$ guarantees that the above estimate vanishes (in fact exponentially fast) in n, thus establishing the claim. □

As before, getting a nearly spanning cycle **whp** can be easily done through sprinkling.

A similar statement holds for the probability space $D(n,p)$ of random directed graphs, with an essentially identical proof.

4 The appearance of Hamilton cycles in random graphs

The main aim of this section is to establish the Hamiltonicity threshold in the probability space $G(n,p)$, this is the minimum value of the edge probability $p(n)$, for which a random graph G drawn from $G(n,p)$ is **whp** Hamiltonian. By doing so we will prove a classical result of Komlós and Szemerédi [11] and independently of Bollobás [12].

Let us start by providing some intuition on where this threshold is expected to be located. It is well known that the threshold probability for connectivity in $G(n,p)$ is $p = \frac{\ln n}{n}$. More explicitly, one can prove (and we leave this as an exercise) that for any function $\omega(n)$ tending to infinity arbitrarily slowly with n, if $p = \frac{\ln n - \omega(n)}{n}$, then **whp** $G \sim G(n,p)$ is not connected, whereas for $p = \frac{\ln n + \omega(n)}{n}$ **whp** $G \sim G(n,p)$ is connected. Perhaps more importantly, the main reason for the threshold for connectivity to be around $\ln n / n$ is that precisely at this value of probability the last isolated vertex in $G(n,p)$ typically ceases to exist. Of course, the graph cannot be connected while having isolated vertices, and this is the easy part of the connectivity threshold statement; the hard(er) part is to prove that if $p(n)$ is such that $\delta(G) \geq 1$ **whp**, then G is **whp** connected.

If so, we can suspect that the threshold for Hamiltonicity of $G(n,p)$ coincides with that of non-existence of vertices of degree at most one, the latter being an obvious necessary condition for Hamiltonicity. This is exactly what was proven in [12] and [11]. Let us therefore set our goal by stating first a fairly accessible result about the threshold for $\delta(G) \geq 2$, both in $G(n,p)$ and in $G(n,m)$.

Proposition 4.1 *Let* $\omega(n)$ *be any function tending to infinity arbitrarily slowly with* n. *Then:*

- *in the probability space* $G(n,p)$,
    1. *if* $p(n) = \frac{\ln n + \ln \ln n - \omega(n)}{n}$, *then* $G \sim G(n,p)$ **whp** *satisfies* $\delta(G) \leq 1$;
    2. *if* $p(n) = \frac{\ln n + \ln \ln n + \omega(n)}{n}$, *then* $G \sim G(n,p)$ **whp** *satisfies* $\delta(G) \geq 2$;
- *in the probability space* $G(n,m)$,
    1. *if* $m(n) = \frac{(\ln n + \ln \ln n - \omega(n))n}{2}$, *then* $G \sim G(n,m)$ **whp** *satisfies* $\delta(G) \leq 1$;
    2. *if* $m(n) = \frac{(\ln n + \ln \ln n + \omega(n))n}{2}$, *then* $G \sim G(n,m)$ **whp** *satisfies* $\delta(G) \geq 2$.

Proof Straightforward application of the first (for proving $\delta(G) \geq 2$) and the second (for proving $\delta(G) \leq 1$) moment methods in both probability spaces; left as an exercise. $\qquad\square$

Hence our goal will be to prove that for $p(n) = \frac{\ln n + \ln \ln n + \omega(n)}{n}$ and for $m(n) = \frac{(\ln n + \ln \ln n + \omega(n))n}{2}$ the random graphs $G(n,p)$ and $G(n,m)$, respectively, are **whp** Hamiltonian.

We will actually prove a stronger, and a much more delicate, result about the hitting time of Hamiltonicity in random graph processes. Let us first define this notion formally. Let $\sigma : E(K_n) \to [N]$ be a permutation of the edges of the complete graph K_n on n vertices, we can write $\sigma = (e_1, \ldots, e_N)$, where $N = \binom{n}{2}$. A *graph process* $\tilde{G} = \tilde{G}(\sigma)$ is a nested sequence $\tilde{G} = (G_i)_{i=0}^N$, where the graph G_i has $[n]$ as its vertex set and $\{e_1, \ldots, e_i\}$, the prefix of σ of length i, as its edge set. The sequence (G_i) thus starts with the empty graph on n vertices, finishes with the complete graph on n vertices, and its i-th element G_i has exactly i edges; moreover, it is nested, as for $i \geq 1$ the graph G_i is obtained from its predecessor G_{i-1} by adding the i-th edge e_i of σ. We can view $\tilde{G}(\sigma)$ as a graph process (as the name indicates suggestively) or as an evolutionary process, unraveling from the empty graph to the complete graph, as guided by σ.

Now, we introduce the element of randomness in the above definition. Suppose the permutation σ is drawn uniformly at random from the set of all $N!$ permutations of the edges of K_n. Then the corresponding process $\tilde{G}(\sigma)$ is called a *random graph process*. We can describe it in the following equivalent way: start by setting G_0 to be the empty graph on n vertices, and for each $1 \leq i \leq N$, obtain G_i by choosing an edge e_i of K_n missing in G_{i-1} uniformly at random and adding it to G_{i-1}. This very nice and natural probability space models a random evolutionary process in graphs; here too we proceed from the empty graph to the complete graph, but in a random fashion.

Random graph processes are important not just because they model evolution very nicely, in fact, they embed the probability spaces $G(n,m)$ for various m; due to standard connections between $G(n,m)$ and $G(n,p)$ one can also claim they "contain" $G(n,p)$ as well. Observe that running a random process $\tilde{G}$ and stopping it (or taking a *snapshot*) at time m produces the probability distribution $G(n,m)$. Indeed, every graph G with vertex set $[n]$ and exactly m edges is the m-th element of the same number of graph processes, namely, of $m!(N-m)!$ of them. Thus, understanding random graph processes usually leads to immediate consequences for $G(n,m)$, and then for $G(n,p)$, and Hamiltonicity is not exceptional in this sense.

Let $\mathcal{P}$ be a property of graphs on n vertices; assume that P is monotone increasing (i.e. adding edges preserves it), and that the complete graph K_n possesses $\mathcal{P}$ (one can think of $\mathcal{P}$ as being the property of Hamiltonicity). Then, given a permutation $\sigma : E(K_n) \to [N]$ and the corresponding graph process $\tilde{G}(\sigma)$, we can define the first moment i when the i-th element G_i of $\tilde{G}$ has $\mathcal{P}$. This is the so-called *hitting time* of $\mathcal{P}$, denoted by $\tau_\mathcal{P}(\tilde{G}(\sigma))$:

$$\tau_\mathcal{P}(\tilde{G}(\sigma)) = \min\{i \geq 0 : G_i \text{ has } \mathcal{P}\}.$$

Of course, due to the monotonicity of $\mathcal{P}$ from this point until the end of the process, the graphs G_i all have $\mathcal{P}$. When $\tilde{G}$ is a random graph process, the hitting time $\tau_\mathcal{P}(\tilde{G})$ becomes a random variable, and one can study its typical behavior. A related task is to compare two hitting times, and to try to bundle them, deterministically or probabilistically.

We now state the main result of this section, due to Ajtai et al. [13] and to Bollobás [12].

Theorem 4.2 *Let $\tilde{G}$ be a random graph process on n vertices. Denote by $\tau_2(\tilde{G})$ and $\tau_\mathcal{H}(\tilde{G})$ the hitting times of the properties of having minimum degree at least 2, and of Hamiltonicity, respectively. Then* **whp***:*

$$\tau_2(\tilde{G}) = \tau_\mathcal{H}(\tilde{G}).$$

In words, for a typical graph process, Hamiltonicity arrives *exactly* at the very moment the last vertex of degree less than 2 disappears. Of course, it cannot arrive earlier deterministically, so the main point of the above theorem is to prove that typically it does not arrive later either.

As we indicated above, random graph process results are usually more powerful than those for concrete random graph models. Here too we are able to derive the results for $G(n,m)$ and $G(n,p)$ easily from the above theorem.

Corollary 4.3 *Let $m(n) = \frac{(\ln n + \ln \ln n + \omega(n))n}{2}$. Then a random graph $G \sim G(n,m)$ is* **whp** *Hamiltonian.*

Proof Generate a random graph G distributed according to $G(n,m)$ by running a random graph process $\tilde{G}$ and stopping it at time m. By Proposition 4.1 we know that $\tau_2(\tilde{G}) \leq m$. Theorem 4.2 implies that typically $\tau_2(\tilde{G}) = \tau_\mathcal{H}(\tilde{G})$, and thus the graph of the process has become Hamiltonian not later than m. Hence G is **whp** Hamiltonian as well. $\qquad\square$

Corollary 4.4 *Let* $p(n) = \frac{\ln n + \ln \ln n + \omega(n)}{n}$. *Then a random graph* $G \sim G(n,m)$ *is* **whp** *Hamiltonian.*

Proof Observe that generating a random graph $G \sim G(n,p)$ and conditioning on its number of edges being exactly equal to m produces the distribution $G(n,m)$. Let $\omega_1(n) = \omega(n)/3$. Denote $I = [Np - n\omega_1(n), Np + n\omega_1(n)]$. Observe that for every $m \in I$, the random graph $G \sim G(n,m)$ is **whp** Hamiltonian by Corollary 4.3. Also, the number of edges in $G(n,p)$ is distributed binomially with parameters N and p and has thus standard deviation less than $\sqrt{Np} \ll n\omega_1(n)$. Applying Chebyshev we derive that **whp** $|E(G)| \in I$. Hence,

$Pr[G \sim G(n,p)$ is not Hamiltonian$]$

$$= \sum_{m=0}^{N} Pr[|E(G)| = m] \cdot Pr[G \text{ is not Hamiltonian}||E(G) = m]$$

$$\leq Pr[|E(G)| \notin I]$$

$$+ \sum_{m \in I} Pr[|E(G)| = m] Pr[G \text{ is not Hamiltonian}| \, |E(G) = m]$$

$$= o(1) + \sum_{m \in I} Pr[|E(G)| = m] Pr[G \sim G(n,m) \text{ is not Hamiltonian}]$$

$$= o(1) \cdot Pr[Bin(N,p) \in I]$$

$$= o(1).$$

$\square$

Now we start proving Theorem 4.2. The proof is somewhat technical, so before diving into its details, we outline its main idea briefly. Recall that our goal is to prove that for a typical random process $\tilde{G}$, we have $\tau_2(\tilde{G}) = \tau_{\mathcal{H}}(\tilde{G})$. In order to prove this, we will take a very close look at the snapshot G_{τ_2} of $\tilde{G}$, aiming to prove that this graph is **whp** Hamiltonian. By definition, the minimum degree of G_{τ_2} is exactly two, and it is thus quite reasonable to expect that this graph is typically a $(k,2)$-expander for $k = \Theta(n)$. This is true indeed; however, such expansion by itself does not quite guarantee Hamiltonicity. As indicated in Section 2.3, expanders form a very convenient backbone for augmenting a graph to a Hamiltonian one – according to Corollary 2.8 every connected non-Hamiltonian $(k,2)$-expander has $\Omega(k^2)$ boosters. Observe though that since we aim to prove a hitting time result, we cannot allow ourselves to sprinkle a few random edges on top of our expander – a Hamilton cycle should appear at the very moment Np the minimum degree in the random graph process becomes two. We will circumvent this difficulty in the following way: we will argue that the snapshot G_{τ_2} typically is not only a good expander by itself, but also contains a subgraph Γ_0 that is about as good an expander as G_{τ_2} is, but contains only a small positive proportion of its edges. Having obtained such Γ_0, we will start looking for boosters relative to Γ_0, but already contained in our graph G_{τ_2} – thus avoiding the need for sprinkling. We will argue that G_{τ_2} is typically such that it contains a booster with respect to every sparse expander in it. If this is the case, then we will be able to start with Γ_0 and update it sequentially by adding a booster after a booster (at most n boosters will need to be added by definition), until we finally reach Hamiltonicity – all within G_{τ_2}. Observe crucially that at each step of this augmentation procedure the updated backbone Γ_i, obtained by adjoining to Γ_0 the previously added boosters, has at most n more edges than Γ_0 and is thus still a sparse subgraph of G_{τ_2}; of course the required expansion is inherited from iteration to iteration. Then our claim about G_{τ_2} typically containing a booster with respect to every sparse expander within is applicable, and we can push the process through. This is quite a peculiar proof idea – it appears that the random graph is helping itself to become Hamiltonian!

Let us get to work. As outlined before, we run a random graph process $\tilde{G}$ and take a snapshot at the hitting time $\tau_2 = \tau_2(\tilde{G})$. Denote

$$m_1 = \frac{n \ln n}{2}$$
$$m_2 = n \ln n.$$

Observe that by Proposition 4.1 we have that **whp** $m_1 \leq \tau_2 \leq m_2$. Let

$$d_0 = \lfloor \delta_0 \ln n \rfloor$$

where $\delta_0 > 0$ is a sufficiently small constant to be chosen later, and denote, for a graph G on n vertices,

$$SMALL(G) = \{v \in V(G) : d(v) < d_0\}.$$

Observe that for $G \sim G(n,m)$ with $m \geq m_1$, the expected vertex degree is asymptotically equal to $\ln n$. Thus falling into $SMALL(G)$ is a rather rare event, and we can expect the vertices of $SMALL(G)$ to be few and far apart in the graph. In addition, such G should typically have a very nice edge distribution, with no small and dense vertex subsets, and many edges crossing between any two large disjoint subsets. This is formalized in the following lemma.

Lemma 4.5 *Let $\tilde{G} = (G_i)_{i=0}^{N}$ be a random graph process on n vertices. Denote $G = G_{\tau_2}$, where $\tau_2 = \tau_2(\tilde{G})$ is the hitting time for having minimum degree 2 in $\tilde{G}$. Then* **whp** *G has the following properties:*

(P1) $\Delta(G) \leq 10 \ln n$; $\delta(G) \geq 2$;

(P2) $|SMALL(G)| \leq n^{0.3}$;

(P3) *G does not contain a nonempty path of length at most 4 such that both of its (possibly identical) endpoints lie in $SMALL(G)$;*

(P4) *every vertex subset $U \subset [n]$ of size $|U| \le \frac{n}{\ln^{1/2} n}$ spans at most $|U| \cdot \ln^{3/4} n$ edges in G;*

(P5) *for every pair of disjoint vertex subsets U, W of sizes $|U| \le \frac{n}{\ln^{1/2} n}$, $|W| \le |U| \cdot \ln^{1/4} n$, the number of edges of G crossing between U and W is at most $\frac{d_0 |U|}{2}$;*

(P6) *for every pair of disjoint vertex subsets U, W of size $|U| = |W| = \left\lceil \frac{n}{\ln^{1/2} n} \right\rceil$, G has at least $0.5n$ edges between U and W.*

Proof The proof is basically a fairly standard (though tedious) manipulation with binomial coefficients. We will thus prove several of the above items, leaving the proof of remaining ones to the reader.

(P1): Observe that since **whp** $\tau_2 \le m_2$, it is enough to prove that in $G \sim G(n, m_2)$ there are **whp** no vertices of degree at least $10 \ln n$. For a given vertex $v \in [n]$, the probability that v has degree at least $10 \ln n$ in $G(n, m_2)$ is at most

$$\binom{n-1}{10 \ln n} \frac{\binom{N - 10 \ln n}{m_2 - 10 \ln n}}{\binom{N}{m_2}} \le \left(\frac{en}{10 \ln n} \right)^{10 \ln n} \left(\frac{m_2}{N} \right)^{10 \ln n},$$

by the standard estimates on binomial coefficients stated in Section 2.1.2. After cancellations we see that the above estimate is at most $(en/5(n-1))^{10 \ln n} = o(1/n)$. Applying the union bound we obtain that typically at time m_2, and thus at $\tau_2 \le m_2$ as well, there are no vertices of degree at least $10 \ln n$. The bound on $\delta(G)$ is immediate from the definition of τ_2.

(P2): Notice that since adding edges can only decrease the size of $SMALL(G)$, it is enough to prove that typically already at time m_1 $|SMALL(G_{m_1})| \le n^{0.3}$. Let $G \sim G(n, m_1)$. If $|SMALL(G)| \ge n^{0.3}$, then G contains a subset $V_0 \subset V$, $|V_0| = k = \lceil n^{0.3} \rceil$ such that $e_G(V_0, V - V_0) \le d_0 k$. The probability of this to happen in $G(n, m_1)$ is at most:

$$\binom{n}{k} \sum_{i \le d_0 k} \binom{k(n-k)}{i} \cdot \frac{\binom{N-k(n-k)}{m_1 - i}}{\binom{N}{m_1}} \le \binom{n}{k} \sum_{i \le d_0 k} \binom{kn}{i} \cdot \frac{\binom{N-k(n-k)}{m_1 - i}}{\binom{N-i}{m_1 - i}} \cdot \frac{\binom{N-i}{m_1 - i}}{\binom{N}{m_1}}$$

$$\le \left(\frac{en}{k} \right)^k \sum_{i \le d_0 k} \left(\frac{ekn}{i} \right)^i \cdot e^{-\frac{(m_1 - i)(k(n-k)-i)}{N-i}} \cdot \left(\frac{m_1}{N} \right)^i$$

$$\le \left(\frac{en}{k} \right)^k \sum_{i \le d_0 k} \left(\frac{ekm_1 n}{iN} \right)^i \cdot e^{-\frac{0.9 k m_1 n}{N}}$$

$$\le \left(\frac{en}{k} \right)^k \cdot (d_0 k + 1) \cdot \left(\frac{ekm_1 n}{d_0 kN} \right)^{d_0 k} \cdot e^{-\frac{0.9 k m_1 n}{N}}$$

$$\le (d_0 k + 1) \left[3n^{0.7} \left(\frac{3 \ln n}{d_0} \right)^{d_0} \cdot e^{-0.8 \ln n} \right]^k = o(1),$$

for δ_0 small enough.

(P3): Since **whp** $m_1 \leq \tau_2 \leq m_2$, it is enough to prove the following statement: **whp** every two (possibly identical) vertices of $SMALL(G_{m_1})$ are not connected by a path of length at most 4 in G_{m_2}.

Let us prove first that **whp** there is no such path in $G_{m_1} \sim G(n, m_1)$. We start with the case where the endpoints of the path are distinct. Fix $1 \leq r \leq 4$, a sequence P of distinct vertices $v_0, \ldots, v_r$ in $[n]$ and denote by $\mathcal{A}_P$ the event $(v_i, v_{i+1}) \in E(G_{m_1})$ for every $0 \leq i \leq r-1$. Then

$$Pr[\mathcal{A}_P] = \frac{\binom{N-r}{m_1-r}}{\binom{N}{m_1}} \leq \left(\frac{m_1}{N}\right)^r = \left(\frac{\ln n}{n-1}\right)^r .$$

If we now condition on $\mathcal{A}_P$, then the two edges (v_0, v_1) and (v_{r-1}, v_r) are present in G_{m_1}. Thus in order for both v_0, v_r to fall into $SMALL(G_{m_1})$, out of $2n-4$ potential edges between $\{v_0, v_r\}$ and the rest of the graph (the edges $(v_0, v_1), (v_{r-1}, v_r)$ are excluded from the count), only at most $2d_0 - 2$ are present in G_{m_1}. Hence:

$$Pr[v_0, v_r \in SMALL(G_{m_1}) | \mathcal{A}_P] \leq \sum_{i=0}^{2d_0-2} \binom{2n-4}{i} \cdot \frac{\binom{N-r-2n+4}{m_1-r-i}}{\binom{N-r}{m_1-r}}$$

$$\leq (2d_0-1)\binom{2n-4}{2d_0-2} \cdot \frac{\binom{N-r-2n+4}{m_1-r-2d_0+2}}{\binom{N-r}{m_1-r}}$$

$$\leq 2d_0\binom{2n-4}{2d_0-2} \cdot \frac{\binom{N-r-2n+4}{m_1-r-2d_0+2}}{\binom{N-r-2d_0+2}{m_1-r-2d_0+2}} \cdot \frac{\binom{N-r-2d_0+2}{m_1-r-2d_0+2}}{\binom{N-r}{m_1-r}}$$

$$\leq 2d_0 \cdot \left(\frac{en}{d_0-1}\right)^{2d_0-2} \cdot e^{-\frac{(m_1-r-2d_0+2)(2n-2d_0-2)}{N-r-2d_0+2}}$$

$$\cdot \left(\frac{m_1-r}{N-r}\right)^{2d_0-2}$$

$$\leq 2d_0 \cdot \left(\frac{em_1 n}{(d_0-1)N}\right)^{2d_0-2} \cdot e^{-\frac{1.9m_1 n}{N}} \leq n^{-1.8} ,$$

for δ_0 small enough. Hence, applying the union bound over all such sequences of $r+1$ vertices, we conclude that the probability that there exists a path in G_{m_1} of length at most 4, connecting two distinct vertices from $SMALL(G_{m_1})$ is at most $\sum_{r \leq 4} n^{r+1} \cdot \left(\frac{\ln n}{n-1}\right)^r \cdot n^{-1.8} = o(1)$. The case where the endpoints of the path are identical is treated similarly.

In light of the above, we can assume that after m_1 steps of the random graph process the current graph does not have a forbidden short path between the vertices of $SMALL$. Moreover, by **(P2)** we can assume that $|SMALL(G_{m_1})| \leq$

$n^{0.3}$. Now, let us run the process between m_1 and m_2. In order for the i-th edge of the process, $m_1 < i \leq m_2$, to close a short path between the vertices of $SMALL(G_{m_1})$, it should fall inside a current set U of vertices at distance at most 3 from $SMALL(G_{m_1})$. We have proven (property **(P1)**) that in fact **whp** the maximum degree of G_{m_2} as well is at most $10 \ln n$. Hence, **whp** in this time interval, the set U has size at most $|SMALL(G_{m_1})| \cdot (10 \ln n)^3$, and thus the probability of the i-th edge of the process to fall inside U is at most $\frac{\binom{|U|}{2}}{N - m_2} = o(n^{-1.3})$. Taking the union bound over all such i in the interval $(m_1, m_2]$, we establish the desired property.

Properties **(P4)**–**(P6)** can be proven quite similarly (and are in fact simpler to prove), and we spare the reader from the (perhaps somewhat boring...) task of reading their proofs.

$\square$

The above stated properties **(P1)**–**(P6)** are sufficient to prove that G_{τ_2} is a very good expander by itself. Our goal is somewhat different though – we aim to prove that G_{τ_2} contains a much sparser, but still fairly good expander. For this purpose, assume that a graph $G = (V, E)$ has properties **(P1)**–**(P6)**. Form a random subgraph Γ_0 of G as follows. For every $v \in V - SMALL(G)$, choose a set $E(v)$ of d_0 edges of G incident to v uniformly at random; for every $v \in SMALL(G)$, define $E(v)$ to be the set of *all* edges of G touching v. Finally, define Γ_0 to be the spanning subgraph of G, whose edge set is:

$$E(\Gamma_0) = \bigcup_v E(v).$$

In words, in order to form Γ_0 we retain all edges touching the vertices of $SMALL(G)$, and sparsify randomly other edges.

Lemma 4.6 *With high probability (over the choices of $E(v)$) the subgraph Γ_0 is a $(k, 2)$-expander with at most $d_0 n$ edges, where $k = \frac{n}{4}$.*

Proof Since by definition $|E(v)| \leq d_0$ for every $v \in V$, it follows immediately that $|E(\Gamma_0)| \leq d_0 n$. We now prove that typically Γ_0 has the following property:

(P7) For every pair of disjoint sets U, W of size $|U| = |W| = \left\lceil \frac{n}{\ln^{1/2} n} \right\rceil$, Γ_0 has at least one edge between U and W.

Fix sets U, W as above. We know by **(P6)** that G has at least $0.5n$ edges between U and W. For a vertex $u \in U$, the probability that none of the edges between u and W falls into $E(u)$ is at most

$$\frac{\binom{d_G(u) - d_G(u,W)}{d_0}}{\binom{d_G(u)}{d_0}} \leq e^{-\frac{d_0 \cdot d_G(u,W)}{d_G(u)}} \leq e^{-\frac{d_0}{10 \ln n} \cdot d_G(u,W)}$$

by **(P1)**. Hence the probability that none of the vertices u from U chooses an edge between u and W to be put into its set $E(u)$ is at most:

$$\prod_{u \in U} e^{-\frac{d_0}{10 \ln n} \cdot d_G(u,W)} = e^{-\frac{d_0}{10 \ln n} \cdot e_G(U,W)} = e^{-\Theta(n)}.$$

Applying the union bound over all choices of U, W gives the desired claim.

We now claim that for every graph G, and every subgraph $\Gamma_0 \subseteq G$ of minimum degree 2 satisfying properties **(P2)**, **(P3)**, **(P4)**, **(P5)**, **(P7)**, the subgraph Γ_0 is an $(n/4, 2)$-expander. In order to verify this claim, let $S \subset [n]$ be a subset of size $|S| \leq n/4$. Denote $S_1 = S \cap SMALL(G)$, $S_2 = S - SMALL(G)$. Consider first the case where $|S_2| \leq \frac{n}{\ln^{1/2} n}$. Since $\delta(\Gamma_0) \geq 2$, and all vertices from $SMALL$ are at a distance of more than 4 from each other by **(P3)**, we obtain: $|N_{\Gamma_0}(S_1)| \geq 2|S_1|$. As for vertices from S_2, they are all of degree at least d_0 in Γ_0. The set S_2 spans at most $|S_2| \cdot \ln^{3/4} n$ edges in G, and thus in Γ_0, according to **(P4)**. It thus follows that $e_{\Gamma_0}(S_2, V - S_2) \geq d_0|S_2| - 2e_{\Gamma_0}(S_2) > \frac{d_0|S_2|}{2}$. Hence $|N_{\Gamma_0}(S_2)| \geq |S_2| \cdot \ln^{1/4} n$, by **(P5)**. Finally, notice that, due to the non-existence of short paths connecting $SMALL(G)$ again, the set $S_1 \cup N_{\Gamma_0}(S_1)$ contains only one vertex from $u \cup N_{\Gamma_0}(u)$ for every $u \in S_2$ (here we use the fact that the forbidden paths have length at most 4). Therefore, $|(S_1 \cup N_{\Gamma_0}(S_1)) \cap (S_2 \cup N_{\Gamma_0}(S_2))| \leq |S_2|$. Altogether,

$$|N_{\Gamma_0}(S)| = |N_{\Gamma_0}(S_2) \setminus S_1| + |N_{\Gamma_0}(S_1) - (S_2 \cup N_{\Gamma_0}(S_2))|$$

$$\geq |S_2|(\ln^{1/4} n - 1) + 2|S_1| - |S_2| \geq 2(|S_1| + |S_2|) = 2|S|,$$

as required. The complementary case $\frac{n}{\ln^{1/2} n} \leq |S_2| \leq \frac{n}{4}$ is very simple: by property **(P7)**, such S_2 misses at most $n/\ln^{1/2} n$ vertices in its neighborhood in Γ_0, also $|S_1| \leq |SMALL(G)| \leq n^{0.3}$ by **(P2)**. It follows that $|N_{\Gamma_0}(S)| \geq n - \frac{n}{\ln^{1/2} n} - |S_2| - |SMALL(G)| \geq \frac{n}{2}$. $\qquad\square$

Notice that every $(\frac{n}{4}, 2)$-expander Γ on n vertices is necessarily connected. Indeed, if such Γ is not connected, then consider its connected component C of size $|C| \leq \frac{n}{2}$, and take U to be an arbitrary subset of C of size $|U| = \min\{\lfloor \frac{n}{4} \rfloor, |C|\}$. Then the external neighborhood of U in Γ has size at least $2|U| > |C - U|$ by our expansion assumption, and falls entirely within C – a contradiction.

As we have stated already in this text, expanders are not necessarily Hamiltonian themselves, but they are amenable to reaching Hamiltonicity by adding extra (random) edges, as they contain many boosters. However, in our circumstances we do not have extra time for sprinkling, and the required boosters should come from within the already existing edges of the random graph. Fortunately, a random graph $G(n,m)$ with $m = m(n)$ in the relevant range has **whp** a booster with respect to any sparse expander it contains, as given by the following lemma.

Lemma 4.7 *Let $\tilde{G} = (G_i)_{i=0}^{N}$ be a random graph process on n vertices. Denote $G = G_{\tau_2}$, where $\tau_2 = \tau_2(\tilde{G})$ is the hitting time for having minimum degree two in $\tilde{G}$. Assume the constant δ_0 is small enough. Then* **whp** *for every $(n/4, 2)$-expander $\Gamma \subset G$ with $V(\Gamma) = V(G)$ and $|E(\Gamma)| \le d_0 n + n$, Γ is Hamiltonian, or G contains at least one booster with respect to Γ.*

Proof Recall that every connected $(k, 2)$-expander Γ is Hamiltonian or has at least $k^2/2$ boosters, by Corollary 2.8. In order for a random graph G to violate the assertion of the lemma, G should contain some $(n/4, 2)$-expander Γ with few edges, but none of at least as many as $n^2/32$ boosters relative to Γ (note that the required connectivity of G is delivered by the expansion of Γ, as explained above). Since we cannot pinpoint the exact location of τ_2, we instead take the union bound over all $m_1 \le m \le m_2$, as **whp** τ_2 is located in this interval. So the estimate is:

$$\sum_{m=m_1}^{m_2} \sum_{i \le d_0 n + n} \frac{\binom{N}{i} \cdot \binom{N - i - \frac{n^2}{32}}{m - i}}{\binom{N}{m}} + o(1) \qquad (4.1)$$

(we sum over all relevant values of m, adding $o(1)$ in the end to account for the probability that τ_2 falls outside the interval $[m_1, m_2]$; then we sum over all possible values i of $|E(\Gamma)|$ and we bound from above by $\binom{N}{i}$ the number of $(n/4, 2)$-expanders with i edges in the complete graph on n vertices, and finally we require the edges of Γ to be present in $G(n, m)$, but all at least $n^2/32$ boosters relative to Γ to be omitted). The ratio of the binomial coefficients above can be estimated as follows:

$$\frac{\binom{N - i - \frac{n^2}{32}}{m - i}}{\binom{N}{m}} \le \frac{e^{-\frac{n^2}{32} \frac{(m-i)}{N-i}} \binom{N-i}{m-i}}{\binom{N}{m}} \le e^{-\frac{m}{17}} \left(\frac{m}{N}\right)^i,$$

assuming that δ_0 in the definition of d_0 is small enough. We can thus estimate the i-th summand in (4.1) as follows:

$$\binom{N}{i} e^{-\frac{m}{17}} \left(\frac{m}{N}\right)^i \le \left(\frac{eN}{i} \cdot \frac{m}{N}\right)^i \cdot e^{-\frac{m}{17}} = \left(\frac{em}{i}\right)^i \cdot e^{-\frac{m}{17}}$$

$$\le \left(\frac{em}{d_0 n + n}\right)^{d_0 n + n} \cdot e^{-\frac{m}{17}} = o(n^{-3}),$$

again for δ_0 small enough (it is even exponentially, and not just polynomially, small in n). Summing over all $i \le d_0 n + n$ and then over all $m_1 \le m \le m_2$ establishes the required claim. $\square$

The stage is now set to deliver the final punch of the proof of Theorem 4.2. Recall that our goal is to prove that for a random graph process $\tilde{G}$, **whp** the graph $G = G_{\tau_2}$ at the very moment τ_2 when the minimum degree becomes 2 is already Hamiltonian. First, observe that by Lemma 4.6 **whp** G contains an $(n/4, 2)$-expander Γ_0 with at most $d_0 n$ edges. We start with this sparse expander Γ_0 and keep adding boosters to it until the current graph Γ_i becomes Hamiltonian; obviously at most n steps (edge additions) will be needed to reach Hamiltonicity. If we ever get stuck before reaching Hamiltonicity, say at step $i \geq 0$, then the current graph Γ_i is still an $(n/4, 2)$-expander, is connected and non-Hamiltonian, and has at most $d_0 n + n$ edges, but the graph G has no boosters with respect to Γ_i. This however does not happen typically due to Lemma 4.7. If so, the process of edge addition eventually completes with a subgraph $\Gamma_i \subset G$, which is Hamiltonian. The proof is complete!

References

[1] A. Frieze and M. Karoński, *Introduction to Random Graphs*, Cambridge University Press, Cambridge, 2016.

[2] B. Bollobás, *Random Graphs*, 2nd edn., Cambridge University Press, Cambridge, 2001.

[3] S. Janson, T.Łuczak, and A. Ruciński, *Random Graphs*, Wiley, New York, 2000.

[4] I. Ben-Eliezer, M. Krivelevich, and B. Sudakov, The size Ramsey number of a directed path, *J. Comb. Theor. Ser. B* **102** (2012), 743–755.

[5] L. Pósa, Hamiltonian circuits in random graphs, *Discrete Math.* **14** (1976), 359–364.

[6] P. Erdős, and A. Rényi, On the evolution of random graphs, *Publ. Math. Inst. Hungar. Acad. Sci.* **5** (1960), 17–61.

[7] M. Ajtai, J. Komlós, and E. Szemerédi, The longest path in a random graph, *Combinatorica* **1** (1981), 1–12.

[8] M. Krivelevich and B. Sudakov, The phase transition in random graphs – a simple proof, *Random Structures Algorithms* **43** (2013), 131–138.

[9] R. Karp, The transitive closure of a random digraph, *Random Structures Algorithms* **1** (1990), 73–93.

[10] W. Fernandez de la Vega, Long paths in random graphs, *Studia Sci. Math. Hungar.* **14** (1979), 335–340.

[11] J. Komlós and E. Szemerédi, Limit distributions for the existence of Hamilton circuits in a random graph, *Discrete Math.* **43** (1983), 55–63.

[12] B. Bollobás, The evolution of sparse graphs, *Graph Theory and Combinatorics*, Academic Press, London (1984), 35–57.

[13] M. Ajtai, J. Komlós, and E. Szemerédi, First occurrence of Hamilton cycles in random graphs, Cycles in graphs (Burnaby, B.C., 1982), North-Holland Mathematical Studies 115, North-Holland, Amsterdam (1985), 173–178.

2

Random graphs from restricted classes

Konstantinos Panagiotou

1 Introduction

Suppose we are given a class $\mathcal{G}$ of graphs. We write

$$\mathcal{G} = \bigcup_{n \geq 1} \mathcal{G}_n,$$

where $\mathcal{G}_n$ contains all graphs in $\mathcal{G}$ that have exactly n vertices. We are interested in the following (informal) question.

Question. Consider a *random graph* G_n from the set $\mathcal{G}_n$. What does it *typically* look like?

A prominent instance of this question is the case where G_n is the *Erdős–Rényi random graph*, typically denoted by $G_{n,p}$. Here any two (distinct) vertices are connected by an edge independently with a probability $p \in [0, 1]$. This model is well studied and by the time of this writing we have a plethora of beautiful results exposing many important properties, see for example the excellent books Bollobás [1] and Janson et al. [2], and also Chapter 1 in this book.

In this chapter we will consider other classes of graphs $\mathcal{G}$ and distributions on $\mathcal{G}_n$ that are rather different from the Erdős–Rényi setting, for example:

 (i) Trees equipped with the uniform distribution.
 (ii) *Outerplanar* graphs, that is, graphs that can be embedded in the plane in a crossing-free way such that there is a face containing all vertices. Equivalently, outerplanar graphs are characterized by not containing the complete graph K_4 nor the complete bipartite graph $K_{2,3}$ as a minor. (Recall that a graph H is said to be a *minor* of G, if H can be constructed from G by deletion of vertices and edges, and by contraction of edges.)
 (iii) *Series–parallel* graphs, which do not contain K_4 as minor.
 (iv) Several families of *planar* graphs, among which (i)–(iii) are particular examples.

Our main focus will be on the typical *global structure* and some *local properties* (for example the distribution of the degree sequence) of such families of graphs.

The graph classes that we consider here admit a so-called *decomposition*, which is a recursive description with the help of general-purpose combinatorial constructions, see Section 3. These constructions have appeared frequently in modern systematic approaches to asymptotic enumeration and random sampling of several combinatorial structures. It is beyond the scope of the present text to present a detailed survey, and apart from the material presented here we refer the reader to Flajolet and Sedgewick [3] and Duchon et al. [4] and the references therein for an excellent introduction and a detailed exposition.

The main contribution in this part of the chapter is a *methodological* one. One benefit of the knowledge of the decomposition of the class under consideration is that it allows us to develop *mechanically* algorithms that sample graphs by using the framework of *Boltzmann samplers*. This framework was introduced in Duchon et al. [4] and it was extended in several directions, see for example the papers by Flajolet et al. [5] and Bodirsky et al. [6]. Indeed, as we will demonstrate in Section 2 for the case of random trees and in Section 3 for the case of richer graph classes, it is possible to relate *systematically* properties of random graphs from the class in question to properties of sequences of *independent* random variables. This enables us to exploit the wealth of powerful theorems and properties of such objects, such as the central and local limit theorem or the Chernoff bounds, in order to study directly properties of random graphs in which the events that certain edges are present or absent are far from being independent.

To give a glimpse into the results that we are going to prove with almost elementary means, let T_n be a uniform random tree with n vertices and let $D_k(T_n)$ be the number of vertices with degree k in T_n. We will show that for $k \in \mathbb{N}$ and any $\alpha \in (0, (k-1)^{-1})$, as $n \to \infty$

$$\Pr\left(D_k(T_n) = \lfloor \alpha n \rfloor\right) \sim C(k, \alpha) n^{-1/2} e^{-nI(k, \alpha)} \tag{1.1}$$

for some $C(k, \alpha) > 0$ and an explicitly given function $I(k, \alpha)$, which actually is the sum of the rate functions of a binomial and a Poisson distribution, see Theorem 2.5. In other words, we will obtain a local limit theorem for the number of vertices of a given degree in T_n. Such results can probably also be shown by different methods (although this has not been done yet), for example with the help of generating functions and complex analytic methods, but the presented method is simple and leads fairly quickly to the estimate (1.1). Apart from that, in Section 2 the presentation of the material is self-contained and no prior knowledge of the area (other than basic probability theory and combinatorics) is required to verify all details.

Our tools will also enable us to study global properties of random graphs from classes that are significantly more complex than trees, for example planar graphs. In Section 3 we are going to demonstrate that a single numerical parameter, which is defined by the number of certain graphs in the class C under consideration, determines the *global* shape of a uniform random graph C_n from C with n vertices, see Theorem 3.9. In particular, there is in a well-defined sense a *dichotomy* regarding the size of the largest 2-connected subgraph of C_n, where in one case it is of logarithmic size, while in the other case it contains $\Theta(n)$ vertices. This behavior resembles very much the evolution of the binomial random graph, where suddenly a giant connected component appears. This structural dichotomy will allow us in some cases to treat systematically several parameters in a unified way for whole families of graph classes.

Acknowledgement. The author of this chapter expresses his gratitude to Nikolaos Fountoulakis and Danny Hefetz for organizing the school "Random Graphs, Geometry and Asymptotic Structure" in Birmingham in August 2013. Moreover, the author thanks Elisabeta Candelero for composing the lecture notes on which this text is partially based, and Benedikt Stufler for many helpful discussions.

1.1 Preliminaries

In this section we collect some estimates that we will use frequently. Without further reference, we will exploit *Stirling's formula*, which states that as $n \to \infty$

$$n! \sim \sqrt{2\pi n}\,(n/e)^n \cdot$$

for the approximation of the factorial function. Actually, the right-hand side of this equation is a lower bound for $n!$, that is, $n! \geq \sqrt{2\pi n}\,(n/e)^n$ for all $n \in \mathbb{N}$.

In our computations we will often need bounds for the probability that a sum of independent Bernoulli distributed random variables deviates from its mean. We will use the following version of the well-known Chernoff bounds, see for example Alon and Spencer [7].

Theorem 1.1 (Chernoff bounds). *Let $X_1, \ldots, X_n$ be independent random variables that take values in $\{0,1\}$, and let $S_n = X_1 + \cdots + X_n$. Then, for all $t \geq 0$*

$$\Pr(S_n \geq \mathbb{E}[S_n] + t) \leq \exp\left\{ -\frac{t^2}{2(\mathbb{E}[S_n] + t/3)} \right\},$$

as well as

$$\Pr(S_n \leq \mathbb{E}[S_n] - t) \leq \exp\left\{ -\frac{t^2}{2\mathbb{E}[S_n]} \right\}.$$

The Chernoff bounds are also valid when the X_i's are Poisson distributed random variables (instead of Bernoulli distributed as in the previous theorem), see for example Alon and Spencer [7]. We shall also make use of this fact a few times.

Let ϕ and Φ be the standard normal density and distribution functions, that is,

$$\phi(x) = (2\pi)^{-1/2} e^{-x^2/2} \quad \text{and} \quad \Phi(x) = (2\pi)^{-1/2} \int_{-\infty}^{x} e^{-t^2/2} dx.$$

We will use the following general result, which enables us to compute asymptotically the probability that the sum of independent random variables attains a specific value. A (rather simple) proof can be found in Davis and McDonald [8].

Theorem 1.2 (Local Limit Theorem). *Let $X_1, X_2, \ldots$ be independent integer-valued random variables, and let $S_n = X_1 + \cdots + X_n$. For $i \in \mathbb{N}$ set*

$$q_i = \sum_{k \in \mathbb{Z}} \min\{\Pr(X_i = k), \Pr(X_i = k+1)\}$$

and let $Q_n = \sum_{i=1}^{n} q_i$. Suppose that S_n satisfies a central limit theorem, that is, for all $x \in \mathbb{R}$

$$\lim_{n \to \infty} \Pr(S_n - a_n < x b_n) = \Phi(x)$$

for some sequences a_n and $b_n > 0$ such that $\lim_{n \to \infty} b_n = \infty$ and $\limsup_{n \to \infty} b_n^2 / Q_n < \infty$. Then S_n satisfies a local limit theorem, that is

$$\lim_{n \to \infty} \sup_{k \in \mathbb{Z}} |b_n \Pr(S_n = k) - \phi((k - a_n)/b_n)| = 0.$$

In our intended applications we will typically use $k = \mathbb{E}[S_n] + O(1)$, where this theorem yields, as $n \to \infty$

$$\Pr(S_n = \mathbb{E}[S_n] + O(1)) \sim \frac{\phi(0)}{b_n} = \frac{1}{b_n \sqrt{2\pi}}.$$

Moreover, a further consequence of the previous theorem (that can also be shown directly with the use of Stirling's formula) is the following well-known property of the binomial distribution sharpening in this specific setting the bounds in Theorem 1.1. Let $S_n = X_1 + \cdots + X_n$, where the X_is are independent and identical Bernoulli distributed random variables with mean $p \in (0, 1)$. Let $\alpha \in (0, 1)$. Then, as $n \to \infty$

$$\Pr(S_n = \lfloor \alpha n \rfloor) \sim (2\pi \alpha (1 - \alpha) n)^{-1/2} e^{n \cdot I_p(\alpha)}, \tag{1.2}$$

where

$$I_p(\alpha) = \alpha \log(p/\alpha) + (1 - \alpha) \log((1 - p)/(1 - \alpha)).$$

This function is known as the *rate function* of the binomial distribution.

2 Random trees

In this section we write $\mathcal{T}_n$ for the set of trees with n vertices and $\mathcal{T} = \cup_{n \geq 1} \mathcal{T}_n$ for the set of all trees. We will be interested in several statistics of T_n, which is a tree drawn uniformly at random among all trees in $\mathcal{T}_n$. The well-known Cayley's formula states that

$$|\mathcal{T}_n| = n^{n-2}.$$

We will make use of this fact several times. For any tree $T \in \mathcal{T}$ and any vertex v in T we write $d_T(v)$ for the degree, that is, the number of neighbors, of v in T. Moreover, for $k \in \mathbb{N}$ we set

$$D_k(T) = |\{v \text{ vertex of } T : d_T(v) = k\}|.$$

Our main objective in this section is to study the distribution of D_k when T is the random tree T_n, and the main results are collected in Theorems 2.2 and 2.5. In order to do so we will proceed with the following three steps, which will be a simple instantiation of the more general method that we will use in the next section:

1. Provide an *algorithm* that generates T_n or something that is "close" to T_n. This algorithm takes as input an infinite sequence of independent random values and constructs a tree out of them.
2. Relate the statistics in question, in this case the distribution of D_k, to the values used by the algorithm.
3. Use properties of sequences of independent random variables to gain information about the distribution of the relevant statistics.

The details of this (abstract) procedure are exposed in the next subsections. In particular, in Section 2.1 we give an algorithm that maps a sequence of independent Poisson random variables $Z = (Z_1, Z_2, \ldots)$ in a simple and natural way to a tree. As we shall see, this does not always result in a tree with exactly n vertices; but when it does so – and this is not a very unlikely event – we show that the result is actually uniformly distributed among all trees in $\mathcal{T}_n$. We also show a simple relation between $D_k(\mathsf{T})$ and the sequence Z. This enables us to study the tails of the distribution of $D_k(\mathsf{T}_n)$ in Section 2.2, where we prove for any $1 \leq k \leq n$ tail estimates that match the Chernoff bounds, see Theorem 2.2. In Section 2.3 we take the approach one step further, and we even show a local limit theorem for $D_k(\mathsf{T}_n)$, that is we compute the probablity of the event "$D_k(\mathsf{T}_n) = \lfloor \alpha n \rfloor$" for any $\alpha \in (0, (k-1)^{-1})$, see Theorem 2.5. The material in this section is self-contained, and no prior knownledge of the area is required to verify all details.

    Random trees and their properties have been the topic of study in many works, and we shall not review them here. An excellent introduction to the topic presenting a variety of results and methods based on the analysis of generating functions is the monograph by Drmota [9]. Moreover, the tour-de-force survey paper by Janson [10] studies systematically several properties of random trees with a probabilistic approach and collects and improves in a unified context many previous results. Both these works are rich sources of references. However, the main theorems presented here are new at the time of this writing, and the focus is on developing a method that is based on studying a suitable stochastic process that generates the tree.

2.1 Random trees via a sampling algorithm

Our investigation of random trees begins with the following construction. Define the set of *rooted trees* on n vertices by

$$\mathcal{T}_n^\bullet := \mathcal{T}_n \times [n],$$

where $[n] = \{1, \ldots, n\}$. A rooted tree is thus a pair (T, v), where T is a tree and v is a distinguished vertex, which we call its root. We write $\mathcal{T}^\bullet = \cup_{n \geq 1} \mathcal{T}_n^\bullet$ for the class of all rooted trees. Consider an element $\mathsf{T}_n^\bullet$ chosen uniformly at random from $\mathcal{T}_n^\bullet$. Since every tree with n vertices corresponds to n distinct rooted trees, we have that in distribution

$$D_k(\mathsf{T}_n^\bullet) \overset{(d)}{=} D_k(\mathsf{T}_n).$$

A similar statement is actually true for any random variable that is not affected by rooting the tree. In the sequel we will thus study the distribution of $D_k(\mathsf{T}_n^\bullet)$.

Our algorithm for generating random rooted trees is a simple stochastic process that first creates the root of the tree and then adds recursively subtrees to it. More concretely, we proceed in the following steps:

1. Create a single "untouched" vertex, which will be the root of the resulting tree.
2. Perform the following actions:
   - Select any untouched vertex (arbitrarily), denote it by v.
   - Given a value n_v distributed like Po(1), that is, according to an independent Poisson random variable of parameter 1, create n_v new untouched vertices.
   - Connect the new vertices to v and declare v "touched."
   Repeat Step 2 *as long as there are untouched vertices.*
3. Let V be the set of all vertices that were created. Label the elements of V uniformly at random with labels from the set $[|V|]$.

Steps 1 and 2 of this process are nothing more than a *branching process* or *Galton–Watson random tree* whose offspring distribution follows the Poisson law with expectation one.

We denote the outcome of the process by $T^\bullet$, and we write $T^\bullet = \infty$ if the output is not defined, that is, the process did not terminate. Note that if $T^\bullet \neq \infty$, then we have $T^\bullet \in \mathcal{T}_n^\bullet$ for some $n \in \mathbb{N}$. Three observations are immediate:

– Note that we have a priori *no control* over the total number of vertices $|T^\bullet|$ of $T^\bullet$, that is, the number of vertices in the generated tree is a random variable.
– All choices (that is, the choice of the n_v's and the permutation of the labels) made in the algorithm are independent random variables.
– Let u be any vertex in $T^\bullet$. Then, by construction

$$d_{T^\bullet}(v) = \begin{cases} n_v, & v \text{ is the root of } T^\bullet, \\ n_v + 1, & \text{otherwise} \end{cases} . \tag{2.1}$$

However, regarding the first observation, our algorithm has the following properties.

Lemma 2.1. *Let $n \in \mathbb{N}$. Then the following statements are true.*

(a) As $n \to \infty$ we have that $\Pr(T^\bullet \in \mathcal{T}_n^\bullet) \sim (2\pi)^{-1/2} n^{-3/2}$.
(b) Let $T^\bullet \in \mathcal{T}_n^\bullet$. Then

$$\Pr(T^\bullet = T^\bullet \mid T^\bullet \in \mathcal{T}_n^\bullet) = |\mathcal{T}_n^\bullet|^{-1} = n^{-n+1}.$$

In other words, conditional on the event "$T^\bullet \in \mathcal{T}_n^\bullet$" we get the uniform distribution on $\mathcal{T}_n^\bullet$, that is $T_n^\bullet$. Hence in the forthcoming proofs we can exchange $T_n^\bullet$ for $T^\bullet$ conditional on having exactly n vertices. Morever, the probability that actually $T^\bullet \in \mathcal{T}_n^\bullet$ is not "too small," that is proportional to an inverse polynomial in n. This will also be used heavily in the forthcoming arguments.

Proof (Lemma 2.1). Let $T^\bullet \in \mathcal{T}_n^\bullet$. We will show that

$$\mathbb{P}(T^\bullet = T^\bullet) = \frac{e^{-n}}{n!}, \tag{2.2}$$

from which the claims follow rather straightorwardly. To see (a) note that by Cayley's formula we have that $|\mathcal{T}_n^\bullet| = n^{n-1}$. Together with Stirling's formula $n! \sim \sqrt{2\pi n}\,(n/e)^n$

$$\Pr(T^\bullet \in \mathcal{T}_n^\bullet) = |\mathcal{T}_n^\bullet| \cdot \frac{e^{-n}}{n!} \sim n^{n-1} \cdot e^{-n} (2\pi n)^{-1/2} (e/n)^n \tag{2.3}$$

and (a) follows. To see (b), note that by Bayes's rule and the first step in (2.3) and (2.2)

$$\Pr(T^\bullet = T^\bullet \mid T^\bullet \in \mathcal{T}_n^\bullet) = \frac{\Pr(T^\bullet = T^\bullet)}{\Pr(T^\bullet \in \mathcal{T}_n^\bullet)} = |\mathcal{T}_n^\bullet|^{-1},$$

as claimed.

In order to show (2.2) we proceed by induction on n. The base case $n = 1$ is established immediately, since $|T_1^\bullet| = 1$ and $\mathsf{T}^\bullet = T^\bullet$ if and only if the first Poisson random variable that is selected in Step 2 of the algorithm equals 0 (that is, the root has no children). Since this happens with probability e^{-1}, (2.2) follows.

Suppose that $n \geq 2$. Note that our algorithm can be formulated equivalently in the following way:

1′ Create a single "untouched" vertex r.
2′ Perform the following actions.
   – Let $D \sim \mathrm{Po}(1)$.
   – Let $\mathsf{T}_1^\bullet, \ldots, \mathsf{T}_D^\bullet$ be D independent rooted trees distributed like $\mathsf{T}^\bullet$.
   – Connect the roots of $\mathsf{T}_1^\bullet, \ldots, \mathsf{T}_D^\bullet$ to r and declare r as the root of the resulting tree.
3′ Let V be the set consisting of r and all vertices in $\mathsf{T}_1^\bullet, \ldots, \mathsf{T}_D^\bullet$. Partition $[|V|]$ uniformly at random into $1 + D$ sets $\mathsf{L}_0, \ldots, \mathsf{L}_D$, where $|\mathsf{L}_0| = 1$ and $|\mathsf{L}_i| = |\mathsf{T}_i^\bullet|$ for $1 \leq i \leq D$. Assign to r the unique label in L_0, and for each $i \in [D]$ to $\mathsf{T}_i^\bullet$ the labels in L_i canonically, i.e. the j-th vertex in $\mathsf{T}_i^\bullet$, $1 \leq j \leq |\mathsf{T}_i^\bullet|$, is assigned the j-th largest label in L_i.

The equivalence of 1–3 to 1′–3′ can easily be established by unfolding the recurrence in 2′; instead of creating each subtree hanging from the root step by step we collect all steps into one "recursive" call to $\mathsf{T}^\bullet$. Moreover, since in the $\mathsf{T}_i^\bullet$'s the labels are assigned randomly in Step 3, it is enough in Step 3′ to just partition the labels in V.

With this observation at hand note that we can associate to the rooted tree $T^\bullet = (T, v)$ a vector $(v, d, \{(T_i^\bullet, L_i)\}_{1 \leq i \leq d})$, where $v \in [n]$ is the root, d is the degree of v and the set $\{(T_i^\bullet, L_i)\}_{1 \leq i \leq d}$ where the $T_i^\bullet$'s are rooted trees and the L_i's form a partition of $[n] \setminus \{v\}$, is obtained as follows. We first remove v from $T^\bullet$. This leaves us with a forest with d trees $T_1, \ldots, T_d$ where the vertices have labels in $[n] \setminus \{v\}$; let L_i be the labels in T_i. Next we identify in each T_i its root as the unique vertex $v_i \in L_i$ that was connected in $T^\bullet$ to the root v. Finally we relabel the vertices in T_i so that they have labels from the set $[|T_i|]$ in such a way that the original labeling can be uniquely reconstructed: the canonical way of doing this is to assign the label 1 to the smallest vertex in T_i, the label 2 to the second smallest, and so on. This relabeling then yields the d rooted trees $T_1^\bullet, \ldots, T_d^\bullet$. With this notation we infer that $\mathsf{T}^\bullet = T^\bullet$ if and only if the following events occur:

- In Step 2′ we have that $D = d$.
- There is a permutation $\pi : [d] \to [d]$ such that for all $1 \leq i \leq d$ we have in Step 2′ that $\mathsf{T}_i^\bullet = T_{\pi(i)}^\bullet$.
- In Step 3′ we have that $\mathsf{L}_0 = \{v\}$, and for all $1 \leq i \leq d$ we have $\mathsf{L}_i = L_{\pi(i)}$.

After fixing π, all the above events are independent. Moreover, the choices of the $T_i^{\bullet}$'s are independent, and the probability of getting the correct partition in Step 3' equals

$$\binom{|T^{\bullet}|}{1,|L_{\pi(1)}|,\ldots,|L_{\pi(d)}|}^{-1} = \binom{n}{1,|L_1|,\ldots,|L_d|}^{-1}.$$

By applying the induction hypothesis

$$\mathbb{P}(T^{\bullet} = T) = \sum_{\substack{\pi \text{ permuta-} \\ \text{tion of } [d]}} \Pr(\mathrm{Po}(1)=d) \cdot \prod_{i\in[d]} \Pr(T_i^{\bullet} = T_i^{\bullet}) \cdot \frac{1}{\binom{n}{1,|L_1|,\ldots,|L_d|}}$$

$$= d! \cdot \frac{e^{-1}}{d!} \cdot \prod_{i\in[d]} \frac{e^{-|T_i^{\bullet}|}}{|T_i^{\bullet}|!} \cdot \frac{\prod_{i\in[d]} |L_i|!}{n!}.$$

Since $|L_i| = |T_i^{\bullet}|$ and $|T_1^{\bullet}| + \cdots + |T_d^{\bullet}| = n - 1$ the claim follows. $\qquad\square$

In the following subsections we will exploit the structure of $T^{\bullet}$ and in particular Equation (2.1) to prove concentration results and limit theorems for the number of vertices of a given degree in the uniform rooted tree $T_n^{\bullet}$. In particular, with Lemma 2.1 it is possible to "exchange" $T_n^{\bullet}$ with $T^{\bullet}$, conditional on $T^{\bullet} \in \mathcal{T}_n^{\bullet}$. That is, Lemma 2.1 asserts for any event $\mathcal{E} \subseteq \mathcal{T}_n^{\bullet}$ that

$$\Pr(T_n^{\bullet} \in \mathcal{E}) = \Pr(T^{\bullet} \in \mathcal{E} \mid T^{\bullet} \in \mathcal{T}_n^{\bullet}).$$

We will exploit this principle several times.

2.2 Coarse estimates

The stochastic process creating the random rooted tree $T^{\bullet}$ can be described alternatively as follows. Let $Z_1, Z_2, \ldots$ be an (infinite) sequence of independent random variables, each one distributed like a Poisson random variable with mean one. Then the algorithm in Section 2.1 for generating $T^{\bullet}$ can be interpreted as follows: it takes as *input* the sequence $Z = (Z_i)_{i\geq 1}$ and every time the value n_v is needed in Step 2, the algorithm takes the next unused value in Z.

Note that if $T^{\bullet}$ has n vertices, then exactly the first n values in Z were used: every vertex is declared untouched when created, and as soon as its children are created, it becomes touched and stays so forever. Moreover, unless a vertex in $T^{\bullet}$ is the root, then by (2.1) its degree equals one plus the value of the corresponding entry in Z. Thus

$$D_k(T^{\bullet}) = \mathbf{1}[Z_1 = k] + \sum_{i=2}^{|T^{\bullet}|} \mathbf{1}[Z_i = k-1]. \tag{2.4}$$

By definition of the Z_i's we have for any $d \in \mathbb{N}_0$

$$\mathbb{P}(Z_i = d) = \frac{e^{-1}}{d!} =: \mu_d.$$

Let $n \in \mathbb{N}$. Note that the random variable

$$\mathbf{1}[Z_1 = k] + \sum_{i=2}^{n} \mathbf{1}[Z_i = k-1]$$

is a sum of *independent* indicator random variables and has expectation

$$M(n,k) = \mu_k + (n-1)\mu_{k-1} = e^{-1}\left(\frac{1}{k!} + \frac{n-1}{(k-1)!}\right). \qquad (2.5)$$

Thus, by the Chernoff bounds we obtain for any $t \geq 0$

$$\Pr\left(\left|\left(\mathbf{1}[Z_1 = k] + \sum_{i=2}^{n} \mathbf{1}[Z_i = k-1]\right) - M(n,k)\right| \geq t\right)$$

$$\leq 2\exp\left\{-\frac{t^2}{2(M(n,k)+t/3)}\right\}. \qquad (2.6)$$

In view of (2.4) such a statement should also hold for a random rooted tree $\mathsf{T}_n^\bullet$, since by Lemma 2.1 the conditioning on $\mathsf{T}^\bullet \in \mathcal{T}_n^\bullet$ does not seem too "severe." Our first result asserts that this is indeed (almost) the case, and in the proof we reduce explicitly the computation of the probability of an event involving D_k to an event resembling the one in (2.6).

Theorem 2.2. *There is a constant $C > 0$ such that the following is true. Let $n, k \in \mathbb{N}$ and $t \geq 0$. Then with $M(n,k)$ as in (2.5)*

$$\Pr\left(|D_k(\mathsf{T}_n^\bullet) - M(n,k)| \geq t\right) \leq Cn^{3/2}\exp\left\{-\frac{t^2}{2(M(n,k)+t/3)}\right\}.$$

Proof. As a shorthand for the notation, let us write $\mathcal{E}^\bullet = \mathcal{E}(n,k,t)$ for the event "$|D_k(\mathsf{T}_n^\bullet) - M(n,k)| \geq t$" and $\mathcal{E}$ for the event "$|D_k(\mathsf{T}^\bullet) - M(n,k)| \geq t$"; we want to estimate the probability of $\mathcal{E}^\bullet$. By Lemma 2.1 the distributions of $\mathsf{T}_n^\bullet$ and $\mathsf{T}^\bullet$, conditional on $\mathsf{T}^\bullet \in \mathcal{T}_n^\bullet$, are the same. Thus

$$\Pr(\mathcal{E}^\bullet) = \Pr(\mathcal{E} \mid \mathsf{T}^\bullet \in \mathcal{T}_n^\bullet) = \frac{\Pr(\mathcal{E}, \mathsf{T}^\bullet \in \mathcal{T}_n^\bullet)}{\Pr(\mathsf{T}^\bullet \in \mathcal{T}_n^\bullet)}. \qquad (2.7)$$

By Lemma 2.1 we know that $\Pr(\mathsf{T}^\bullet \in \mathcal{T}_n^\bullet) \sim cn^{-3/2}$, for some $c > 0$. Hence there is a $C' > 0$ such that $\Pr(\mathsf{T}^\bullet \in \mathcal{T}_n^\bullet) \leq C'n^{-3/2}$ for all $n \in \mathbb{N}$. Moreover, using (2.4), the event "$\mathcal{E}, \mathsf{T}^\bullet \in \mathcal{T}_n^\bullet$" implies that

$$\left|\left(\mathbf{1}[Z_1 = k] + \sum_{i=2}^{n} \mathbf{1}[Z_i = k-1]\right) - M(n,k)\right| \geq t.$$

Using (2.7) yields

$$\Pr(\mathcal{E}^\bullet) \leq \frac{n^{3/2}}{C'} \Pr\left(\left|\left(\mathbf{1}[Z_1 = k] + \sum_{i=2}^{n} \mathbf{1}[Z_i = k-1]\right) - M(n,k)\right| \geq t\right).$$

We have thus reduced the study of $\mathcal{E}^\bullet$ to the study of the properties of independent random variables. By applying (2.6) and choosing $C = 2/C'$ the claim follows immediately. $\square$

Note that in the previous theorem there is no restriction whatsoever on n, k, t. However, in order for the bound on the probability to be useful we need that the exponent is at least logarithmic in n, for example

$$\frac{t^2}{2(M(n,k)+t/3)} \geq 2C \log n.$$

This is in particular the case when we choose

$$t = \omega\left((M(n,k)\log n)^{1/2}\right).$$

In view of those estimates Theorem 2.2 can be used for example to derive a lower bound for the *maximum degree* Δ_n in T_n respectively $T_n^\bullet$. Indeed, from (2.5) we know that $M(n,k) = n/e(k-1)! + O(1)$. Set

$$d^*(n) = \max\{d \in \mathbb{N} : (d-1)! \leq n/e\}.$$

With Stirling's formula $d! = (d/e)^{d+O(1)}$ we infer after a slightly tedious but standard calculation that as $n \to \infty$

$$d^*(n) \sim \frac{\log n}{\log \log n}.$$

The following result, whose proof can be found below, is a simple consequence of Theorem 2.2.

Theorem 2.3. *Let $\varepsilon > 0$. With $M(n,k)$ as in (2.5)*

$$\Pr\left(\forall 1 \leq k \leq d^*(n) - 2 : |D_k(T_n^\bullet) - M(n,k)| < \varepsilon M(n,k)\right) = 1 - o(1).$$

Moreover, for all $1 \leq k \leq d^*(n) - 2$ we have that $M(n,k) \geq (\log n)^{2-o(1)}$.

In other words, we obtain that the *family* $(D_k(T_n^\bullet))_{1 \leq k \leq d^*(n)-2}$ of random variables concentrates around the vector $(M(n,k))_{1 \leq k \leq d^*(n)-2}$. In particular, we immediately obtain the following bound on the maximum degree.

Corollary 2.4. $\Pr(\Delta_n \geq d^*(n) - 2) = 1 - o(1)$.

Proof (Theorem 2.3). The definition of $d^*(n)$ implies immediately that for any $1 \leq k \leq d^*(n) - 2$ we have $M(n,k) \geq (\log n)^{2-o(1)}$. Let $\mathcal{B}_k$, $1 \leq k \leq d^*(n) - 2$, be the event that $|D_k(\mathsf{T}_n^\bullet) - M(k,n)| \geq \varepsilon M(k,n)$. With Theorem 2.2 we obtain that there is a $C > 0$ such that

$$\Pr(\mathcal{B}_k) \leq Cn^{3/2}\exp\left\{-\frac{\varepsilon^2 M(n,k)}{2(1+\varepsilon/3)}\right\} = o(1/n).$$

By the union bound, with probability $1 - o(1)$ none of the $\mathcal{B}_k$'s will occur and the statement follows. □

The bound in Theorem 2.3 is not tight, as it can be shown that with high probability we actually have

$$|\Delta_n - d^*(n)| \leq 1$$

and even that Δ_n concentrates on at most two values in this interval (and even more can be shown), see Janson [10] and the references in that paper. However, the strength of the approach presented here is, beside its simplicity, that it yields uniform bounds that hold for arbitrary deviations almost all the way up to the maximum degree.

2.3 Fine estimates

In this section we will improve significantly the (coarse) bounds on the distribution of $D_k(\mathsf{T}_n^\bullet)$ that were derived in Theorem 2.2, with the ultimate aim of showing a local limit theorem. As in Section 2.2 we view the stochastic process creating the random rooted tree $\mathsf{T}^\bullet$ as follows. We let $Z = (Z_1, Z_2, \ldots)$ be an infinite sequence of independent random variables that are distributed like a Poisson random variable with mean one. Then the algorithm in Section 2.1 for generating $\mathsf{T}^\bullet$ can be interpreted in the following way: it takes Z as input, and every time the value n_v in Step 2 is needed, the algorithm takes the next unused value in Z.

To facilitate the presentation we let $D'_k(T^\bullet)$ denote the number of vertices that have *out*degree k in a rooted tree $T^\bullet = (T, v)$; the outdegree of a vertex u is defined as the number of neighbors of u in T that have a distance larger than u from v. Then, by (2.1) and as in (2.4)

$$D'_k(\mathsf{T}^\bullet) = \sum_{i=1}^{|\mathsf{T}^\bullet|} \mathbf{1}[Z_i = k-1]. \tag{2.8}$$

That is, D_k and D'_k differ by at most one. In the sequel we are going to study $D'_k(\mathsf{T}^\bullet)$ instead of $D_k(\mathsf{T}^\bullet)$ because it makes matters a bit technically simpler, but all arguments can be adapted easily to fit to $D_k(\mathsf{T}^\bullet)$ as well. Our main result is the following.

Theorem 2.5. *Let $k \in \mathbb{N}$ and $\alpha \in (0, (k-1)^{-1})$. Then there is a $C = C(k, \alpha) > 0$ such that as $n \to \infty$*

$$\Pr\left(D'_k(\mathsf{T}^\bullet_n) = \lfloor \alpha n \rfloor\right) \sim C n^{-1/2} e^{nI(k,\alpha)}, \qquad (2.9)$$

where

$$I(k, \alpha) = \alpha \log\left(\frac{\mu_{\lambda,k-1}}{\alpha}\right) + (1 - \alpha) \log\left(\frac{1 - \mu_{\lambda,k-1}}{1 - \alpha}\right) + 1 - \lambda + \log \lambda.$$

Here $\mu_{\lambda,k-1} = e^{-\lambda} \lambda^{k-1}/(k-1)!$ is the probability that a Poisson distributed random variable with mean λ equals $k-1$ and λ is the unique positive solution to the equation

$$\frac{\lambda - (k-1)\mu_{\lambda,k-1}}{1 - \mu_{\lambda,k-1}} = \frac{1 - (k-1)\alpha}{1 - \alpha}. \qquad (2.10)$$

From our proof below the value of C in (2.9) can actually be worked out, but we omit the (solely technical) details for simplicity. The proof begins with our usual trick: we exchange $\mathsf{T}^\bullet_n$ with $\mathsf{T}^\bullet$, that is, by applying Lemma 2.1 we get that

$$\Pr\left(D'_k(\mathsf{T}^\bullet_n) = \lfloor \alpha n \rfloor\right) = \Pr\left(D'_k(\mathsf{T}^\bullet) = \lfloor \alpha n \rfloor \mid \mathsf{T}^\bullet \in \mathcal{T}^\bullet_n\right).$$

Our first step is a characterization of the sequences in $Z = (Z_1, Z_2, \ldots)$ that guarantee that $\mathsf{T}^\bullet$ has exactly n vertices. We write $S_n = Z_1 + \cdots + Z_n$ in the rest of this section.

Lemma 2.6. *Let $n \in \mathbb{N}$. Then*

$$\{\mathsf{T}^\bullet \in \mathcal{T}^\bullet_n\} = \{S_n = n - 1, \forall 1 \leq N < n : S_N \geq N\}.$$

Proof. Define the random variables

$$U_i = 2 - i + S_{i-1} = 2 - i + \sum_{j=1}^{i-1} Z_i, \quad i \in \mathbb{N}.$$

Note that U_i equals the number of untouched vertices at the beginning of the ith iteration of Step 2 of the process generating $\mathsf{T}^\bullet$, if there was such an iteration. Indeed, in Step 1 exactly one untouched vertex is created, and thus at the beginning of the first iteration of Step 2 $U_1 = 1$. Moreover, as in the j-th iteration of Step 2 a Poisson number Z_j of new untouched vertices is created, and exactly one untouched vertex is declared touched, we obtain the expression for $U_i, i \geq 2$. It follows that the stocastic process generating $\mathsf{T}^\bullet$ stops after exactly I iterations of Step 2 were performed, where

$$I = \min\{i \in \mathbb{N} : U_i = 0\} - 1.$$

Thus, for all $1 \leq i \leq I$ we have that

$$U_i \geq 1 \Leftrightarrow \sum_{j=1}^{i-1} Z_i \geq i-1.$$

Moreover,

$$U_{I+1} = 0 \Leftrightarrow \sum_{j=1}^{I} Z_i = I-1.$$

Note that I equals the total number of created (initially untouched) vertices, since in every iteration of Step 2 of the process exactly one untouched vertex becomes touched. That is, $I = |\mathsf{T}^\bullet|$ and the claim follows. $\qquad\square$

With this observation at hand, suppose that we want to determine $\Pr(\mathsf{T}_n^\bullet \in \mathcal{P})$, where $\mathcal{P} \subseteq \mathcal{T}_n^\bullet$. Then with Lemma 2.1 and Lemma 2.6 we may proceed as follows:

$$\Pr(\mathsf{T}_n^\bullet \in \mathcal{P}) = \Pr(\mathsf{T}^\bullet \in \mathcal{P} \mid \mathsf{T}^\bullet \in \mathcal{T}_n^\bullet)$$

$$= \Pr(\mathsf{T}^\bullet \in \mathcal{P} \mid S_n = n-1, \forall 1 \leq N < n : S_N \geq N).$$

The conditioning in this expression seems very difficult to handle; the next lemma comes to help in many relevant cases. For a sequence $z = (z_1, \ldots, z_n) \in \mathbb{N}_0^n$ and an index $1 \leq \ell \leq n$ the ℓ-th *rotation* $\rho_\ell(z)$ is defined as

$$\rho_\ell(z) = (z_\ell, \ldots, z_n, z_1, \ldots, z_{\ell-1}).$$

Lemma 2.7. *Let $n \in \mathbb{N}$ and let $\mathcal{Z} \subseteq \mathbb{N}_0^n$. Call $\mathcal{Z}$ rotation invariant if for every $z \in \mathcal{Z}$ all rotations of z are in $\mathcal{Z}$ too. Then, if $\mathcal{Z}$ is rotation invariant,*

$$\Pr\Big((Z_1, \ldots, Z_n) \in \mathcal{Z} \mid S_n = n-1, \forall 1 \leq N < n : S_N \geq N\Big)$$

$$= \Pr\Big((Z_1, \ldots, Z_n) \in \mathcal{Z} \mid S_n = n-1\Big).$$

Proof. Let $\mathcal{Z}'$ contain all those $z \in \mathcal{Z}$ for which additionally $\sum_{1 \leq i \leq n} z_i = n-1$. Note that $\mathcal{Z}'$ is rotation invariant too, and that it is sufficient to show the statement for $\mathcal{Z}'$.

Let us abbreviate $\mathcal{I}_n$ the event "$S_n = n-1, \forall 1 \leq N < n : S_N \geq N$." Given $z \in \mathcal{Z}'$, by the Dvoretzky–Motzkin cycle lemma, see for example Takács [11], there is exactly one rotation $\rho_\ell(z)$ such that for all $1 \leq N < n$ we have that $\sum_{i=1}^{N} (\rho_\ell(z))_i \geq N$. Let $r : \mathcal{Z}' \to \mathcal{Z}'$ be the mapping that assigns to each $z \in \mathcal{Z}'$ this unique rotation of it. Then, if $z \neq r(z)$

$$\Pr\Big((Z_1, \ldots, Z_n) = z \mid \mathcal{I}_n\Big) = 0.$$

Let $\mathcal{Z}''$ contain all those $z \in \mathcal{Z}'$ for which additionally $z = r(z)$. We obtain that

$$\Pr\Big((Z_1, \ldots, Z_n) \in \mathcal{Z}' \mid \mathcal{I}_n\Big) = \Pr\Big((Z_1, \ldots, Z_n) \in \mathcal{Z}'' \mid \mathcal{I}_n\Big). \qquad (2.11)$$

By Bayes's rule

$$\Pr\Big((Z_1,\ldots,Z_n)\in \mathcal{Z}''\mid \mathcal{I}_n\Big)=\frac{\Pr((Z_1,\ldots,Z_n)\in \mathcal{Z}'',\mathcal{I}_n)}{\Pr(\mathcal{I}_n)}.$$

However, the event $(Z_1,\ldots,Z_n)\in \mathcal{Z}''$ implies that $\forall 1\leq N<n:S_N\geq N$, and so the numerator equals $\Pr((Z_1,\ldots,Z_n)\in \mathcal{Z}'',S_n=n-1)$. Moreover, again by the cycle lemma and the fact that for any sequence z and any permutation $\bar{z}$ of it we have that $\Pr((Z_1,\ldots,Z_n)=z)=\Pr((Z_1,\ldots,Z_n)=\bar{z})$ we obtain

$$\Pr(\mathcal{I}_n)=\frac{1}{n}\Pr(S_n=n-1).$$

Combining everything with (2.11) yields with Bayes's rule that

$$\Pr\Big((Z_1,\ldots,Z_n)\in \mathcal{Z}'\mid \mathcal{I}_n\Big)=n\Pr\Big((Z_1,\ldots,Z_n)\in \mathcal{Z}''\mid S_n=n-1\Big).$$

Note that since $\mathcal{Z}'$ is rotation invariant, for any $z\in \mathcal{Z}''$ all n rotations of z (including z) are in $\mathcal{Z}'$. Moreover, for any z and any permutation $\bar{z}$ of it we have

$$\Pr((Z_1,\ldots,Z_n)=z\mid S_n=n-1)=\Pr((Z_1,\ldots,Z_n)=\bar{z}\mid S_n=n-1)$$

and thus

$$\Pr((Z_1,\ldots,Z_n)\in \mathcal{Z}''\mid S_n=n-1)=\frac{\Pr((Z_1,\ldots,Z_n)\in \mathcal{Z}'\mid S_n=n-1)}{n}.$$

The claim follows. $\qquad\qquad\square$

With the previous two lemmas at hand we can study the distribution of the number of vertices of a given outdegree in $\mathsf{T}_n^{\bullet}$.

Proof (Theorem 2.5). We begin by exchanging $\mathsf{T}_n^{\bullet}$ for $\mathsf{T}^{\bullet}$ by applying Lemma 2.1:

$$\Pr\Big(D_k'(\mathsf{T}_n^{\bullet})=\lfloor an\rfloor\Big)=\Pr\Big(D_k'(\mathsf{T}^{\bullet})=\lfloor an\rfloor\mid \mathsf{T}^{\bullet}\in \mathcal{T}_n^{\bullet}\Big).$$

By applying Lemma 2.6 we infer that the conditioning is equivalent to the event "$S_n=n-1,\forall 1\leq N<n:S_N\geq N$," where $S_i=Z_1+\cdots+Z_i$ and the Z_j's are independent Poisson random variables with mean one. Define the event

$$\mathcal{Z}:=\Big\{(z_1,\ldots,z_n)\in \mathbb{N}_0^n:\sum_{1\leq i\leq n}\mathbf{1}[z_i=k-1]=\lfloor an\rfloor\Big\}.$$

Note that $\mathcal{Z}$ is rotation invariant. By using (2.8) and applying Lemma 2.7 we obtain that

$$\Pr\Big(D_k'(\mathsf{T}_n^{\bullet})=\lfloor an\rfloor\Big)=\Pr(\mathcal{Z}\mid S_n=n-1) \qquad (2.12)$$

and further, by Bayes's rule

$$\Pr\Big(D_k'(\mathsf{T}_n^{\bullet})=\lfloor an\rfloor\Big)=\frac{\Pr(S_n=n-1\mid \mathcal{Z})\,\Pr(\mathcal{Z})}{\Pr(S_n=n-1)}.$$

Note that $\Pr(\mathcal{Z})$ is the probability that a binomially distributed random variable takes a specific value (that can be written down explicitly), and that $\Pr(S_n = n - 1)$ is the probability that a $\text{Po}(n)$ random variable equals $n - 1$ (that is also given explicitly). The only troublemaker is the term involving the conditioning on $\mathcal{Z}$. Here we are going to use a simple yet powerful property of the Poisson distribution, so that we can apply the Local Limit Theorem to compute that probability asymptotically. Let $Z'_1, \ldots, Z'_n$ be independent Poisson random variables with mean $\lambda > 0$, and let $S'_n = Z'_1 + \cdots + Z'_n$. Then for any $\mathcal{X} \subseteq \mathbb{N}_0^n$ such that for all $(x_1, \ldots, x_n) \in \mathcal{X}$ we have that $x_1 + \cdots + x_n = s$, where $s \in \mathbb{N}$,

$$\Pr(\mathcal{X} \mid S_n = s) = \Pr(\mathcal{X} \mid S'_n = s), \tag{2.13}$$

that is, we can exchange the Z_i's for the Z'_i's. Note that actually it does not make a difference if $\mathcal{X} \subseteq \mathbb{N}_0^n$ is arbitrary, since $x \in \mathcal{X}$ with $x_1 + \cdots + x_n \neq s$ contribute to neither side of the equation. To see (2.13) we will argue that the right-hand side does not depend on λ. Indeed, by Bayes's rule

$$\Pr(\mathcal{X} \mid S'_n = s) = \frac{\Pr(\mathcal{X})}{\Pr(S'_n = s)} = \frac{\sum_{x \in \mathcal{X}} \prod_{i=1}^n \Pr(Z'_i = x_i)}{\Pr(\text{Po}(\lambda n) = s)}.$$

Using the definition of the Poisson distribution and that for all $x \in \mathcal{X}$ we have $x_1 + \cdots + x_n = s$

$$\Pr(\mathcal{X} \mid S'_n = s) = \frac{\sum_{x \in \mathcal{X}} \prod_{i=1}^n e^{-\lambda} \lambda^{x_i}/x_i!}{e^{-\lambda n}(\lambda n)^s/s!} = \frac{\sum_{x \in \mathcal{X}} \prod_{i=1}^n 1/x_i!}{n^s/s!},$$

an expression that does not depend on λ; this establishes (2.13).

We choose (with foresight) λ as in (2.10). This equation has exactly one solution $\lambda > 0$, by Lemma 2.8 below; the proof of that lemma is self-contained and does not require any other fact proved here. With (2.13) we obtain from (2.12)

$$\Pr\left(D'_k(T^\bullet_n) = \lfloor \alpha n \rfloor\right) = \frac{\Pr(S'_n = n - 1 \mid \mathcal{Z}) \Pr(\mathcal{Z})}{\Pr(S'_n = n - 1)}. \tag{2.14}$$

Let $n'' = n - \lfloor \alpha n \rfloor$ and let $Z''_1, \ldots, Z''_{n''}$ be independent random variables distributed like $\text{Po}(\lambda)$, conditional on being $\neq k - 1$, that is,

$$\Pr(Z''_i = d) = \frac{\mu_{\lambda,d}}{1 - \mu_{\lambda,k-1}} \cdot \mathbf{1}[d \neq k - 1].$$

Let $S''_{n''} = Z''_1 + \cdots + Z''_{n''}$. By conditioning on the set of $\lfloor \alpha n \rfloor$ indices in $[n]$ where the corresponding variables are equal to $k - 1$ it follows from the definition that

$$\Pr(S'_n = n - 1 \mid \mathcal{Z}) = \Pr\left(S''_{n''} = n(1 - (k - 1)\alpha) + O(1)\right).$$

Note that

$$\mathbb{E}[Z''_i] = \frac{\lambda - (k - 1)\mu_{\lambda,k-1}}{1 - \mu_{\lambda,k-1}} \overset{(2.10)}{=} \frac{1 - (k - 1)\alpha}{1 - \alpha}$$

and thus $\mathbb{E}[S''_{n''}] = n(1 - (k-1)\alpha) + O(1)$. Moreover, $S''_{n''}$ satisfies a central limit theorem with expectation and variance linear in n'', being a sum of independent random variables with finite variance $\sigma = \sigma(\alpha,k) > 0$. Since the Z''_i's have support in $\mathbb{N}_0 \setminus \{k-1\}$ the remaining conditions in Theorem 1.2 can be verified easily. Indeed, with the notation there we readily obtain for all $1 \le i \le n''$ that $q_i > 0$ and thus $Q_{n''} = \Theta(n'')$. In particular, the sequence b_n^2/Q_n is bounded and we may infer that there is $c = c(\alpha,k) > 0$ such that as $n \to \infty$

$$\Pr(S'_n = n - 1 \mid \mathcal{Z}) \sim c(\alpha,k) n^{-1/2}.$$

By plugging this into (2.14)

$$\Pr\left(D'_k(\mathsf{T}^\bullet_n) = \lfloor \alpha n \rfloor\right) = c(\alpha,k) n^{-1/2} \cdot \frac{\Pr(\mathcal{Z})}{\Pr(S'_n = n - 1)}. \qquad (2.15)$$

From the definition of $\mathcal{Z}$ we get that

$$\Pr(\mathcal{Z}) = \Pr\left(\mathrm{Bin}(n, \mu_{\lambda,k-1}) = \lfloor \alpha n \rfloor\right),$$

and (1.2) guarantees that as $n \to \infty$

$$\Pr(\mathcal{Z}) \sim (2\pi \alpha(1-\alpha)n)^{-1/2} e^{n \cdot I_{\lambda,k}(\alpha)},$$

where

$$I_{\lambda,k}(\alpha) = \alpha \log\left(\frac{\mu_{\lambda,k-1}}{\alpha}\right) + (1-\alpha)\log\left(\frac{1 - \mu_{\lambda,k-1}}{1 - \alpha}\right).$$

Moreover,

$$\Pr(S'_n = n - 1) = \Pr(\mathrm{Po}(\lambda n) = n - 1) = e^{-\lambda n}\frac{(\lambda n)^{n-1}}{(n-1)!}.$$

With Stirling's formula this is asymptotically

$$\Pr(S'_n = n - 1) \sim (2\pi \lambda^2 n)^{-1/2} \exp\{(1 - \lambda + \log\lambda)n\}$$

and combining all these facts with (2.15) completes the proof.   □

The final ingredient is the following technical lemma regarding the set of solutions to (2.10).

Lemma 2.8. *Suppose that $\alpha \in (0, (k-1)^{-1})$. Then (2.10) has exactly one positive solution.*

Proof. Suppose first that $k \ge 3$ and α are fixed. Note that

$$\frac{\partial}{\partial\lambda}\mu_{\lambda,k-1} = -\mu_{\lambda,k-1} + \mu_{\lambda,k-2}. \qquad (2.16)$$

We can view the left-hand side of (2.10) as a function of λ; its derivative then is

$$\frac{1 - (k - 2 - \lambda)\mu_{\lambda,k-1} - (k - 1 - \lambda)\mu_{\lambda,k-2}}{(1 - \mu_{\lambda,k-1})^2}.$$

We will show that for $\lambda > 0$

$$h_k(\lambda) := (k - 2 - \lambda)\mu_{\lambda,k-1} + (k - 1 - \lambda)\mu_{\lambda,k-2} < 1 \qquad (2.17)$$

from which the claim follows, as then the left-hand side of (2.10) is increasing and $h_k(0) = 0$, while the right-hand side of the same equation is a constant greater than 0. Using the definition of μ, we obtain that

$$h_k(\lambda) = e^{-\lambda} \frac{\lambda^{k-1}}{(k-1)!}(1 + \lambda^{-1}(k - 1 - \lambda)^2).$$

Using Stirling's formula and writing $\lambda = \beta(k - 1)$, where $\beta > 0$, we infer that

$$h_k(\lambda) \le (2\pi(k-1))^{-1/2}\beta^{k-1} e^{(1-\beta)(k-1)}(1 + (1 - \beta)^2\beta^{-1}(k - 1)). \qquad (2.18)$$

We will now investigate the set of λ's for which the right-hand side is maximized. Abbreviating $K = k - 1$, its derivative (with respect to β) is given by

$$(K/2\pi)^{1/2}\beta^{K-2}e^{-(\beta-1)K}(1 - \beta)(K\beta^2 - 2K\beta + K - 1).$$

Setting this to 0 yields $\beta \in \{1, 1 + K^{-1/2}, 1 - K^{-1/2}\}$, and for all these values we infer from (2.18) that $h_k < 1$. Moreover, for $\beta \to \infty$ and $\beta \to 0$ the derivative tends to 0. This completes the proof for $k \ge 3$. The case of $k = 1$ can be shown similarly by noting that (2.16) changes to $\frac{\partial}{\partial\lambda}\mu_{\lambda,k-1} = -e^{-\lambda}$. Finally, for $k = 2$ the right-hand side of (2.10) equals 1, regardless of the value of α, and it follows immediately that $\lambda = 1$ is the only solution. $\qquad \square$

3 Random graphs from block-stable classes

In this section we will study random graphs from classes that are more complex than trees. We will be interested in the behavior of several local and global statistics as the number of vertices in the graph tends to infinity. Our general approach will be along the lines of the methodology described in the previous section: we will develop systematically an algorithm that takes as input an infinite sequence of independent random values and from them generates a random graph with n vertices with a probability that is not too small. Then we will relate the statistics in question to the values used by the algorithm, and we will exploit properties of sequences of independent random variables to gain information about the distribution of the relevant statistics.

Any graph may be decomposed into its connected components, that IS, its maximal connected subgraphs. These connected graphs admit further a so-called *block decomposition* that we recall here briefly. A connected graph is called *2-connected*, if it has at least three vertices, none of which is a cut vertex. A *block* of an arbitrary graph G is a maximal connected subgraph $B \subseteq G$

that does not contain a cut vertex. A well-known fact in graph theory, see for example Diestel [12], asserts that any block is either a 2-connected graph, or an edge, or an isolated vertex. Moreover, any two different blocks intersect in most at one vertex. Here we are interested in classes that are closed under the operation of exchanging any block in any graph by another 2-connected graph in the class.

Definition 3.1. *Let G denote a class of graphs and let $B \subseteq G$ be the set of all graphs that are 2-connected or consist of only two vertices joined by an edge. We say that G is* block-stable, *if $B \neq 0$ and $G \in G$ if and only if every block of G is isomorphic to some graph in B or to a single isolated vertex.*

A few concrete examples of block-stable classes are the following:

- Forests: B consists only of the graph that is a single edge.
- Planar graphs: B consists of all 2-connected planar graphs, together with the graph that is a single edge.
- Outerplanar graphs: B consists of all 2-connected outerplanar graphs (dissections of polygons), together with the graph that is a single edge.
- Cactus graphs: B is the class of all cycles.

Many other prominent classes of graphs are block-stable, for example the class of graphs colorable with a given number of colors, or classes that exclude a list of given 2-connected graphs as subgraphs or minors.

Outline In the following subsections we will study the typical asymptotic structure of a graph that is drawn uniformly at random from a given class of block-stable graphs with n vertices. In Section 3.1 we give a concise introduction of the framework of combinatorial species that will allow us in Section 3.2 to develop systematically a stochastic process that generates graphs from the block-stable class under consideration, in the same spirit as this was performed for trees in previous section. In Section 3.3 we collect some basic asymptotic estimates required for the further study of the sampling algorithm, as for example the probability that a graph with exactly n vertices is generated. In the following two sections, 3.4 and 3.5, we study, through our sampling algorithm, the block distribution in random graphs and we discover a striking dichotomy: depending on a single numerical parameter of the class under consideration, it turns out, see Theorem 3.9, that a typical graph from the class either has a *simple* structure, meaning that all blocks are small (of logarithmic size) or *complex*, that is, there is a giant block that contains a constant fraction of all vertices. We demonstrate in Section 3.6 how the structure of "simple" graphs can be exploited systematically to study other properties, in this case the distribution of the number of vertices with a given degree. The section closes with an overview of other results and references to the literature.

Related and Ongoing Work Random graphs from block-stable classes have been the object of study in many works. The systematic research of such classes was triggered by the paper Giménez and Noy [13], where the number of planar graphs was determined asymptotically. The dichotomy regarding the largest block and higher order components was discovered in Panagiotou and Steger [14] and Fountoulakis and Panagiotou [15], and finer details (for example the limiting distribution in some cases) were described in Giménez et al. [16]. Moreover, there is a wealth of other works studying local and global parameters, as for example the degree distribution, see for example Drmota et al. [17], Bernasconi et al. [18] and Drmota et al. [19], or the global metric defined by such graphs, see Panagiotou et al. [20].

3.1 Combinatorial constructions and Boltzmann samplers

In this section we collect some parts of the theory of combinatorial species and the Boltzmann sampling framework tailored to our specific purpose. *Combinatorial species* allow for a unified treatment of a wide range of objects appearing frequently in modern combinatorial theories. We give only a very concise introduction and refer to Joyal [21] and Bergeron et al. [22] for a detailed discussion and many examples, and to Flajolet and Sedgewick [3] for the development of the equivalent language of *combinatorial classes*.

Combinatorial species A combinatorial species is defined as a family of mappings $\mathcal{F}$ that maps any finite set U of *labels* to a finite set $\mathcal{F}[U]$ of $\mathcal{F}$-*objects* and any bijection $\sigma : U \to V$ to a bijective *transport function* $\mathcal{F}[\sigma] : \mathcal{F}[U] \to \mathcal{F}[V]$, such that

- For all bijections $\sigma : U \to V$, $\sigma' : V \to W$: $\mathcal{F}[\sigma' \circ \sigma] = \mathcal{F}[\sigma'] \circ \mathcal{F}[\sigma]$.
- Let $\mathrm{id}_U : U \to U$ denote the identity map. Then $\mathcal{F}[\mathrm{id}_U] = \mathrm{id}_{\mathcal{F}[U]}$ for all finite sets U.

In the language of category theory, a species is a functor. A simple example is the "species of graphs": it maps each finite U to the set of all graphs with vertex set U, and each bijection $\sigma : U \to V$ naturally induces a bijection from the set of graphs with vertex set U to the set of graphs with vertex set V.

We need a bit more notation regarding species. Let $\mathcal{F}$ and $\mathcal{G}$ be species. We write $\mathcal{G} \subseteq \mathcal{F}$ and say that $\mathcal{G}$ is a *subspecies* of $\mathcal{F}$, if $\mathcal{G}[U] \subseteq \mathcal{F}[U]$ for all finite U and $\mathcal{G}[\sigma] = \mathcal{F}[\sigma]|_{\mathcal{G}[U]}$ for all bijections $\sigma : U \to V$. An example is the "species of trees" as a subspecies of the species of graphs.

Let $\mathcal{F}$ be a species. We say that $\gamma \in \mathcal{F}[U]$ has size $|\gamma| := |U|$ and γ and $\gamma' \in \mathcal{F}[V]$ are termed isomorphic if there is a bijection $\sigma : U \to V$ such that $\mathcal{F}[\sigma](\gamma) = \gamma'$. For $n \in \mathbb{N}_0$ we let $\mathcal{F}_n := \mathcal{F}[\{1, \dots, n\}]$ and by slight abuse of notation we will often identify $\mathcal{F}$ with $\cup_{n \in \mathbb{N}_0} \mathcal{F}_n$. The *exponential generating series* of $\mathcal{F}$ is the formal power series

$$F(x) = \sum_{n \geq 0} |\mathcal{F}_n| \frac{x^n}{n!}.$$

Note that F may have radius of convergence zero. If this is not the case, then we say that $\mathcal{F}$ is *analytic*, and we call F its *exponential generating function (egf)*. The species of (all) graphs is not analytic, since the number of graphs with n vertices is $2^{\binom{n}{2}}$. On the other hand, from Cayley's formula it follows easily that the species of trees is analytic with radius of convergence e^{-1}.

An *isomorphism* $\phi : \mathcal{F} \to \mathcal{G}$ is a family $(\phi_U : \mathcal{F}[U] \to \mathcal{G}[U])_U$ of bijections, where U ranges over all finite sets, such that for all bijections $\sigma : U \to V$ we have that $\mathcal{G}[\sigma] \circ \phi_U = \phi_V \circ \mathcal{F}[\sigma]$. We write $\mathcal{F} \simeq \mathcal{G}$ and say that $\mathcal{F}$ and $\mathcal{G}$ are *isomorphic* if there is an isomorphism in the previous sense. If two species are isomorphic, then the corresponding generating series coincide.

Constructions and generating functions The framework of combinatorial species offers a whole bunch of *constructions* that enable us to create new species from others, and which relate the corresponding generating series; these constructions appear frequently in modern theories of combinatorial analysis and in systematic approaches to random generation of combinatorial objects. In particular, these operations will enable us to develop a recursive description of block-stable classes of graphs, see (3.1) below. We begin with three basic (analytic) species:

- The *empty species* $\mathbf{0}$ with $\mathbf{0}[U] = \emptyset$ for all finite U. Its egf equals 0.
- The *singleton species* $\mathcal{Z}$ with $\mathcal{Z}[U] = \{U\}$ if $|U| = 1$ and $\mathcal{Z}[U] = \emptyset$ otherwise. Its egf equals x. Note that $\mathcal{Z}$ is isomorphic to the subspecies of the species of all graphs that contain only one vertex (and no edge).
- The *set species* SET with SET$[U] = \{U\}$ for all finite sets U. Its egf is given by

$$\sum_{n \geq 0} 1 \cdot \frac{x^n}{n!} = e^x.$$

We continue with three combinatorial operators ("product," "substitution" and "rooting") that may be used to form more complex species out of simpler ones. In the following let $\mathcal{F}$ and $\mathcal{G}$ denote species. The *product* $\mathcal{F} \cdot \mathcal{G}$ is defined by the disjoint union

$$(\mathcal{F} \cdot \mathcal{G})[U] = \bigsqcup_{\substack{(U_1, U_2) \\ \text{partition of } U}} \mathcal{F}[U_1] \times \mathcal{G}[U_2] \quad \text{for all finite sets } U$$

and $(\mathcal{F} \cdot \mathcal{G})[\sigma](f, g) = (\mathcal{F}[\sigma|_{U_1}], \mathcal{G}[\sigma|_{U_2}])$ for all bijections $\sigma : U \to V$ and $(f, g) \in \mathcal{F}[U_1] \times \mathcal{G}[U_2]$. Thus, objects of size n are pairs of $\mathcal{F}$-objects and $\mathcal{G}$-objects whose sizes add up to n. The notation for the product of two species is quite suggestive: from the definition it follows that the exponential generating series for $\mathcal{F} \cdot \mathcal{G}$ is the product $F(x)G(x)$.

If the species $\mathcal{G}$ has no objects of size zero, we can form the *substitution* $\mathcal{F} \circ \mathcal{G}$ by

$$(\mathcal{F} \circ \mathcal{G})[U] = \bigsqcup_{\pi \text{ partition of } U} \mathcal{F}[\pi] \times \prod_{P \in \pi} \mathcal{G}[P] \quad \text{for all finite sets } U.$$

We can interpret an object in $\mathcal{F} \circ \mathcal{G}$ as an $\mathcal{F}$-object whose labels are substituted by objects from $\mathcal{G}$. The transport along a bijection $\sigma : U \to V$ is defined by applying the induced map $\sigma' : \pi \to \pi' := \{\sigma(P) \mid P \in \pi\}$ to the $\mathcal{F}$-object and the maps $\sigma|_P$, $P \in \pi$, to the (corresponding) objects in $\mathcal{G}$. Again, the notation for the substitution is suggestive: from the definition it also follows that the exponential generating series for $\mathcal{F} \circ \mathcal{G}$ equals $(F \circ G)(x) = F(G(x))$.

The *rooted* $\mathcal{F}$-species is defined by

$$\mathcal{F}^{\bullet}[U] = \mathcal{F}[U] \times U \quad \text{for all finite sets } U$$

with componentwise transport. A rooted object is thus created by distinguishing a label, which we call the *root*, and any transport function must preserve roots. This is completely analogous to the rooting of trees that we performed in Section 2.1. The generating series $F^{\bullet}(x)$ of $\mathcal{F}^{\bullet}$ equals $x\frac{d}{dx}F(x)$. Similarly, the *derived* species $\mathcal{F}'$ is given by

$$\mathcal{F}'[U] = \mathcal{F}[U \cup \{*_U\}] \quad \text{for all finite sets } U$$

with $*_U$ referring to an arbitrary fixed element that is not contained in U. The transport along a bijection $\sigma : U \to V$ is defined by $\mathcal{F}'[\sigma](f) = \mathcal{F}[\sigma'](f)$, where σ' preserves the $*$'s, that is, $\sigma'(u) = \sigma(u)$, $u \in U$, and $\sigma'(*_U) = *_V$. Derived objects can be interpreted as canonically rooted objects from $\mathcal{F}$, where the root is always $*_U$. It follows easily that rooted and derived classes are related by an isomorphism $\mathcal{F}^{\bullet} \simeq \mathcal{Z} \cdot \mathcal{F}'$; in particular, the generating series of $\mathcal{F}'$ is $\frac{d}{dx}F(x)$.

Boltzmann samplers The combinatorial constructions presented previously have two important advantages: first, as already demonstrated, they can be used to obtain equations that relate the generating functions of the involved species. Second, as we will demonstrate now, these constructions translate immediately to stochastic processes that generate objects from the given species according to a specific probability measure defined over the whole species, the so-called *Boltzmann model*. This model and the associated randomized algorithms were introduced in the landmark paper by Duchon et al. [4] and were developed further in Flajolet et al. [5] and Fusy [23]. Following these sources we will briefly recall the theory of Boltzmann samplers to the extent required in this section. Let $\mathcal{F} \neq \mathbf{0}$ be an analytic species. Given $x > 0$ such that $0 < F(x) < \infty$, a *Boltzmann sampler* F_x is a random generator such that for any $\gamma \in \mathcal{F} =$

$\cup_{n \geq 0} \mathcal{F}_n$

$$\Pr(\mathsf{F}_x = \gamma) = \frac{x^{|\gamma|}}{F(x)|\gamma|!}.$$

In particular, if we condition on $\mathsf{F}_x \in \mathcal{F}_n$, then we get the uniform distribution on $\mathcal{F}_n$. That is, if we denote by F_n an object drawn uniformly at random from $\mathcal{F}_n$, then for any $\mathcal{E} \subseteq \mathcal{F}$ we have that

$$\Pr(\mathsf{F}_n \in \mathcal{E}) = \Pr(\mathsf{F}_x \in \mathcal{E} \mid \mathsf{F}_x \in \mathcal{F}_n).$$

Note that this is completely analogous to what we did in Section 2.1 in our study of random trees. More specifically, if we let $\mathcal{F} = \mathcal{T}^{\bullet}$ then in the Boltzmann model with parameter $x = e^{-1}$ we get that the probability that a specific $T^{\bullet} \in \mathcal{T}_n^{\bullet}$ is drawn is $e^{-n}/n!$ (since $T^{\bullet}(e^{-1}) = 1$, c.f. Flajolet and Sedgewick [3]). With Lemma 2.1 we immediately obtain the statement that the stochastic process generating $T^{\bullet}$ in Section 2.1 is nothing other than a realization of a Boltzmann sampler with parameter e^{-1} for $\mathcal{T}^{\bullet}$.

We describe Boltzmann samplers for all constructions and basic species that we need using an informal pseudo-code notation in the table below. There we let $\mathrm{Po}(\lambda)$ denote a Poisson distributed generator with mean λ. Note that if $\mathcal{F} = \mathcal{A} \cdot \mathcal{B}$ or $\mathcal{F} = \mathcal{A} \circ \mathcal{B}$ and $0 < F(x) < \infty$, then the samplers for $\mathcal{A}$ and $\mathcal{B}$ are almost surely called with valid parameters, since the coefficients of all power-series involved are non-negative.

Species	Boltzmann Sampler with parameter x
$\mathcal{Z}$	**return** the unique structure of size 1
SET	$m \leftarrow \mathrm{Po}(x)$ **return** the unique structure of size m
$\mathcal{A} \cdot \mathcal{B}$	**return** $(\mathsf{A}_x, \mathsf{B}_x)$ relabeled uniformly at random
$\mathcal{A} \circ \mathcal{B}$	$\gamma \leftarrow \mathsf{A}_y$ with $y = B(x)$ **for** $i = 1$ **to** $|\gamma|$ $\qquad \gamma_i \leftarrow \mathsf{B}_x$ **return** $(\gamma, (\gamma_i)_i)$ relabeled uniformly at random

Finally, suppose that $\mathcal{F}$ and $\mathcal{G}$ are isomorphic species. Then there is a (size preserving) isomorphism Φ between $\mathcal{F}$ and $\mathcal{G}$. If we have a Boltzmann sampler for $\mathcal{G}$, then this can be used to develop a corresponding sampler for $\mathcal{F}$ by applying Φ, that is, $\Phi(\mathsf{G}_x)$ is a Boltzmann sampler with parameter x for $\mathcal{F}$.

3.2 Sampling from block-stable graph classes

Block-stable classes satisfy the following combinatorial specifications that can be found for example in the early work by Joyal [21] and Harary and Palmer [24]:

$$\mathcal{G} \simeq \text{SET} \circ \mathcal{C} \qquad \text{and} \qquad \mathcal{C}^{\bullet} \simeq \mathcal{Z} \cdot (\text{SET} \circ \mathcal{B}' \circ \mathcal{C}^{\bullet}). \tag{3.1}$$

The first isomorphism describes the fact that any graph on a given vertex set can be formed by partitioning it and constructing a connected graph (corresponding to the species $\mathcal{C}$) on each class of the partition. The isomorphism for the class of rooted connected graphs is based on the block decomposition: a rooted connected graph consists of a root vertex r, and is an unordered collection of graphs from $\mathcal{B}'$, which are merged at their roots that are identified with r, and where every vertex different from r is the root of a $\mathcal{C}^{\bullet}$-object. By the rules for computing the generating series we immediately obtain the well-known relations

$$G(x) = \exp(C(x)) \qquad \text{and} \qquad C^{\bullet}(x) = x \exp(B'(C^{\bullet}(x))). \tag{3.2}$$

The following lemma was given in Panagiotou et al. [20].

Lemma 3.2. *Let $\mathcal{C}$ be a block-stable class of connected graphs, $\mathcal{B} \neq 0$ its subclass of all graphs that are 2-connected or a single edge. Then the exponential generating series $C(z)$ has radius of convergence $\rho_{\mathcal{C}} < \infty$ and the sums $y_{\mathcal{C}} := C^{\bullet}(\rho_{\mathcal{C}})$ and $\lambda_{\mathcal{C}} := B'(y_{\mathcal{C}})$ are finite and satisfy*

$$y_{\mathcal{C}} = \rho_{\mathcal{C}} e^{\lambda_{\mathcal{C}}}. \tag{3.3}$$

Moreover, $\mathcal{C}$ is analytic if and only if $\mathcal{B}$ is analytic.

The lemma guarantees that the Boltzmann model for $\mathcal{C}^{\bullet}$ with parameter $0 < x = \rho_{\mathcal{C}}$ is well defined, if $\mathcal{B}$ is analytic. Using the rules for the construction of Boltzmann samplers in Section 3.1 we obtain from the isomorphism in (3.1) the following stochastic process for generating random graphs from $\mathcal{C}^{\bullet}$, which we denote for brevity by $\mathsf{C}^{\bullet}$, according to the Boltzmann model with parameter $\rho_{\mathcal{C}}$:

$\gamma \leftarrow$ a single root vertex
$k \leftarrow \mathsf{Po}(\lambda_{\mathcal{C}})$ $(\star)$
for $i = 1 \ldots k$
    $B \leftarrow \mathsf{B}'_{y_{\mathcal{C}}}$, drop the labels $(\star\star)$
    merge γ with the $*$-vertex of B
    for each non $*$-vertex v of B
        $C \leftarrow \mathsf{C}^{\bullet}$, drop the labels
        merge v with the root of C
return γ relabeled uniformly at random

Note that the above process just reverses the decomposition given in (3.1): it starts with a single vertex, attaches to it a random set of graphs in $\mathcal{B}'$, and proceeds recursively to substitute every newly generated vertex by a rooted connected graph. Note also that this process is a general version of the process described in Section 2.1 for generating the random tree $\mathsf{T}^\bullet$.

For us it will be convenient to replace the sampler stochastic process for $\mathsf{C}^\bullet$ by a slightly different process with the property that the output distributions are the same. Observe that the process makes (among others) two kinds of random choices: first, when it chooses a random value according to a Poisson distribution in the line marked with $(\star)$, and second, when it requires a graph from $\mathcal{B}'$ with parameter y_C in the line marked with $(\star\star)$. We adapt the process by making the random choices *in advance*, and by providing them as part of the input. More precisely, let $K = (K_1, K_2, \ldots)$ be an infinite sequence of non-negative integers, each one chosen independently according to the distribution $\mathrm{Po}(\lambda_C)$, and let $B' = (B'_1, B'_2, \ldots)$ be an infinite sequence of graphs from $\mathcal{B}'$, drawn independently according to the Boltzmann distribution with parameter y_C. Then the process $\mathsf{C}^\bullet(K, B')$, which simulates $\mathsf{C}^\bullet$ by using the next unused value from the provided lists, generates obviously every graph from $\mathcal{C}^\bullet$ with the same probability as $\mathsf{C}^\bullet$. In the sequel we will therefore assume that $\mathsf{C}^\bullet$ in fact denotes the sampler $\mathsf{C}^\bullet(K, B')$ where we often omit the lists K, B' for ease of notation.

As the lists K and B' consist of entries that are independent random variables, the next lemma relating some basic properties of $\mathsf{C}^\bullet$ to properties of K, B' will be a key step in our analysis. For a graph G and $\ell \geq 2$ let $b(\ell; G)$ denote the number of blocks that contain exactly ℓ vertices, and let $b(G) = \sum_{\ell \geq 2} b(\ell; G)$ denote the total number of blocks. Moreover, for a vertex v in G let $c(v; G)$ be its *cut-degree*, that is the number of connected components of G where v is removed.

Lemma 3.3. *Suppose that $\mathsf{C}^\bullet$ was generated using the first n values in K and the first m graphs in B'. Then*

1. $n = |\mathsf{C}^\bullet|$,
2. $m = \sum_{j=1}^{n} K_j$,
3. $m = b(\mathsf{C}^\bullet)$,
4. *For any $\ell \geq 2$ we have that $b(\ell; \mathsf{C}^\bullet) = \sum_{j=1}^{m} \mathbf{1}[|B'_j| = \ell - 1]$.*
5. $\max_{1 \leq v \leq n} c(v; \mathsf{C}^\bullet) = \max\{K_1, \max_{2 \leq j \leq n}(1 + K_j)\}$.

Proof. We discuss the general proof strategy, and refer to Panagiotou and Steger [14] for a complete proof. Recall how $\mathsf{C}^\bullet$ is constructed. Initially, a single vertex r is generated. Then the *number* of graphs in $\mathcal{B}'$ that have r as a common vertex is determined to be K_1. The K_1 graphs $B'_1, \ldots, B'_{K_1}$ are glued together at r. Denote by V the set of non-root vertices in $B'_1, \ldots, B'_{K_1}$. After these steps are performed, the process proceeds recursively for each $v \in V$, where $|V|$ graphs $(C_v)_{v \in V}$ from $C^\bullet$ are generated. Note that for all $v \in V$ we have that $|C_v| < |C^\bullet|$. This calls for a proof by induction on $|C^\bullet|$; all statements are trivially true for the base case $|C^\bullet| = 1$.

We show as an example the induction step for the first statement only. Denote for $v \in V$ by n_v the number of variables from K that the sampler used to generate C_v, and note that due to the induction hypothesis we have $n_v = |C_v|$. But then, as in the construction of γ the root vertex of γ_v is identified with v for all $v \in V$, we have that

$$n = 1 + \sum_{v \in V} n_v = 1 + \sum_{v \in V} |\gamma_v| = |C^\bullet|.$$

$\square$

3.3 Asymptotic analysis

Let C be an analytic block-stable class of connected graphs, $B \subseteq C$ the class consisting of all 2-connected graphs in C and possibly the graph that is a single edge. In Section 3.2 we demonstrated that a stochastic process can be constructed, such that its outcome $C^\bullet$ satisfies for any $C^\bullet \in C^\bullet$

$$\mathbb{P}(C^\bullet = C^\bullet) = \frac{\rho_C^{|C^\bullet|}}{|C^\bullet|! y_C},$$

where $\rho_C > 0$ is the radius of convergence of $C(x)$ (and $C^\bullet(x)$) and $0 < y_C = C^\bullet(\rho_C) < \infty$. Let $c_n^\bullet = |C_n^\bullet|$. Then we get immediately

$$\mathbb{P}(C^\bullet \in C_n^\bullet) = c_n^\bullet \frac{\rho_C^n}{n! y_C}. \tag{3.4}$$

In the case that C is the species T of trees, we computed in this probability Lemma 2.1; in particular, we showed that $\mathbb{P}(T^\bullet \in T_n^\bullet) \sim \Theta(n^{-3/2})$ is only polynomially small in n; this was an important ingredient in all our proofs. In the following we will investigate under what conditions we can expect a similar property to be true for a general block-stable class.

Let us begin with a collection of examples. If C is the species of all connected planar graphs we have by the celebrated result in Giménez and Noy [13]

$$|C_n| \sim c n^{-7/2} \rho_C^{-n} n! \quad \text{and} \quad |B_n| \sim b n^{-7/2} \rho_B^{-n} n!,$$

for some constants $c, b > 0$ and $0 < \rho_B, \rho_C < 1$. In the case of outerplanar (and also series-parallel, cactus, etc.) graphs we have a qualitatively similar behavior, see Drmota et al. [17] and Panagiotou et al. [20], namely

$$|C_n| \sim c n^{-5/2} \rho_C^{-n} n! \quad \text{and} \quad |B_n| \sim b n^{-5/2} \rho_B^{-n} n!,$$

for some constants $c, b > 0$ and $0 < \rho_B, \rho_C < 1$ that depend on the class under consideration only.

Definition 3.4. *Let C be a block-stable and analytic class of connected graphs. We call C nice if there are constants $c, b > 0$ as well as $\rho_C, \rho_B > 0$ and $\alpha, \beta \in \mathbb{R}$ such that*

$$|C_n| \sim c n^{-\alpha} \rho_C^{-n} n! \quad \text{and} \quad |B_n| \sim b n^{-\beta} \rho_B^{-n} n!.$$

By applying Lemma 3.2 we obtain that for a nice class we always have $\alpha > 2$. We obtain from (3.4) immediately the following general statement.

Lemma 3.5. *Let C be nice. Then, as $n \to \infty$, $\mathbb{P}(C^\bullet \in C_n^\bullet) \sim \frac{c}{\gamma_C} n^{-\alpha+1}$.*

3.4 The distribution of blocks

We begin our study by showing Chernoff-type bound for the tails of the distribution of the total number of blocks in random graphs from block-stable classes.

Lemma 3.6. *Let C be a block-stable and analytic class of connected graphs. Then there is a $C > 0$ such that the following is true. For any $n \in \mathbb{N}$ and $t \geq 0$*

$$\Pr(|b(C_n^\bullet) - \lambda_C n| \geq t) \leq C n^{\alpha-1} \exp\left\{ -\frac{t^2}{2(\lambda_C n + t/3)} \right\}.$$

Proof. Using that the distributions of $C_n^\bullet$ and $C^\bullet$, conditional on $C^\bullet \in C_n$ are the same we obtain that

$$\Pr(|b(C_n^\bullet) - \lambda_C n| \geq t) = \Pr(|b(C^\bullet) - \lambda_C n| \geq t \mid C^\bullet \in C_n^\bullet).$$

By appyling Lemma 3.5 and Bayes's rule we obtain that there is a constant $C' > 0$ such that

$$\Pr(|b(C_n^\bullet) - \lambda_C n| \geq t) \leq C' n^{\alpha-1} \Pr(|b(C^\bullet) - \lambda_C n| \geq t, C^\bullet \in C_n^\bullet).$$

By Lemma 3.3, second and third statement, the total number of blocks in $C^\bullet$ equals the sum of the first n values in K, where $n = |C^\bullet|$ by the first statement in Lemma 3.3. Thus, the event "$|b(C^\bullet) - \lambda_C n| \geq t, C^\bullet \in C_n^\bullet$" implies that

$$\left| \sum_{1 \leq j \leq n} K_j - \lambda_C n \right| \geq t,$$

where the K_j's are independent and identically distributed $\mathrm{Po}(\lambda_C)$ random variables. So, by applying the Chernoff bounds we obtain that the desired probability is at most

$$C' n^{\alpha-1} \Pr(|\mathrm{Po}(\lambda_C n) - \lambda_C n| \geq t) \leq 2C' n^{\alpha-1} e^{-t^2/(2\lambda_C n + 2t/3)}.$$

Choosing $C = 2C'$ yields the claimed statement. $\qquad\square$

In particular, we obtain that for any $\varepsilon > 0$ and n sufficiently large that

$$\Pr(|b(C_n^\bullet) - \lambda_C n| \leq \varepsilon n) = 1 - o(1).$$

Note that the proof of Lemma 3.6 is conceptually quite similar to the proof of Theorem 2.2, where we showed Chernoff-type bounds for the tail of the distribution of the number of vertices with a given degree in a random tree. As shown in Section 2.3, with a little more work it is possible to make the estimates much more precise and to obtain a local limit theorem – this program could in principle be performed for $b(C_n^\bullet)$ as well, but we put our focus on different parameters here. Next we provide tail bounds for the number of blocks with a given number of vertices in $C_n^\bullet$.

Lemma 3.7. *There is a $C > 0$ such that the following is true. For any $\ell, n \in \mathbb{N}$ and any $0 < \varepsilon < 1$*

$$\mathbb{P}(|b(\ell; C_n^\bullet) - b_\ell n| \geq \varepsilon b_\ell n) \leq C n^\alpha \left(e^{-\varepsilon^2 b_\ell n/10} + e^{-\varepsilon^2 \lambda_C n/64} \right),$$

where $b_\ell = |\mathcal{B}'_{\ell-1}| y_C^{\ell-1}/(\ell-1)!$.

Proof. Define the event

$$\mathcal{E} = \{C_n^\bullet \in \mathcal{C}_n^\bullet : |b(C_n^\bullet) - \lambda_C n| \leq \varepsilon \lambda_C n/4\}$$

Then, by Lemma 3.6 we get that there is a constant $C' > 0$ such that $\Pr(C_n^\bullet \notin \mathcal{E}) \leq C' n^{\alpha-1} e^{-\varepsilon^2 \lambda_C n/64}$. So,

$$\Pr(|b(\ell, C_n^\bullet) - b_\ell n| \geq \varepsilon b_\ell n) \leq \Pr(|b(\ell; C_n^\bullet) - b_\ell n| \geq \varepsilon b_\ell n, \mathcal{E}) + \Pr(C_n^\bullet \notin \mathcal{E}). \tag{3.5}$$

Using again that $C^\bullet$ conditional on $C^\bullet \in \mathcal{C}_n^\bullet$ has the same distribution as $C_n^\bullet$ we infer that

$$\Pr(|b(\ell; C_n^\bullet) - b_\ell n| \geq \varepsilon b_\ell n, \mathcal{E}) = \Pr(|b(\ell; C^\bullet) - b_\ell n| \geq \varepsilon b_\ell n, \mathcal{E} \mid C^\bullet \in \mathcal{C}_n^\bullet)$$

and Lemma 3.5 guarantees that there is a constant $C'' > 0$ such that this is at most

$$C'' n^{\alpha-1} \Pr(|b(\ell; C^\bullet) - b_\ell n| \geq \varepsilon b_\ell n, \mathcal{E}, C^\bullet \in \mathcal{C}_n^\bullet). \tag{3.6}$$

The above event, together with the fourth statement in Lemma 3.3 implies that for some M satisfying $|M - \lambda_C n| \leq \varepsilon \lambda_C n/4$

$$\left| \sum_{1 \leq j \leq M} \mathbf{1}[|B'_j| = \ell - 1] - b_\ell n \right| \geq \varepsilon b_\ell n.$$

To pin down the right value for b_ℓ, note that with $y_C = C^\bullet(\rho_C)$ and $\lambda_C = B'(y_C)$ and the definition of the Boltzmann model

$$\Pr(|B'_j| = \ell - 1) = |\mathcal{B}'_{\ell-1}| \frac{y_C^{\ell-1}}{(\ell-1)! B'(y_C)} = \frac{|\mathcal{B}_\ell| y_C^{\ell-1}}{(\ell-1)! \lambda_C}, \quad 1 \leq j \leq M,$$

and thus with the definition of b_ℓ,

$$\left| \mathbb{E}\left[\sum_{1 \leq j \leq M} \mathbf{1}[|B'_j| = s] \right] - b_\ell n \right| \leq \varepsilon b_\ell n/4.$$

Putting everything together and applying the Chernoff bounds and the union bound for all M in the given interval we arrive at

$$\Pr(|b(\ell; \mathsf{C}^\bullet) - b_\ell n| \geq \varepsilon b_\ell n, \mathcal{E}, \mathsf{C}^\bullet \in \mathcal{C}_n^\bullet) \leq 2n e^{-\varepsilon^2 b_\ell n/4}.$$

Together with (3.5) and (3.6) this implies that the desired probability is at most

$$\max\{C', 2C''\} n^\alpha \left(e^{-\varepsilon^2 b_\ell n/4} + e^{-\varepsilon^2 \lambda_C n/64} \right)$$

and the claim follows. $\qquad\square$

With this statement at hand we infer that for any $0 < \varepsilon < 1$ and any sequence ℓ such that, say, $\varepsilon^2 b_\ell n \geq 10(\alpha + 2) \log n$, we have that $|b(\ell; \mathsf{C}_n^\bullet) - b_\ell n| \leq \varepsilon b_\ell n$ with probability at least $1 - O(n^{-2})$. Finally, we study the maximal cut-degree $\max_{v \in [n]} c(v; \mathsf{C}_n^\bullet)$ in $\mathsf{C}_n^\bullet$, or equivalently the maximal number of blocks in which any vertex in $\mathsf{C}_n^\bullet$ is contained.

Lemma 3.8. *Let C be a block-stable and analytic class of connected graphs. Then*

$$\Pr\left(\max_{v \in [n]} c(v; \mathsf{C}_n^\bullet) \geq \log n \right) = n^{-\omega(1)}.$$

Proof. We begin with our usual trick and exchange the distributions of $\mathsf{C}_n^\bullet$ and $\mathsf{C}^\bullet$. For a graph G with vertex set $[n]$, let $\mathcal{L}(G)$ be the event that $\max_{v \in [n]} c(v; G) \geq \log n$. Then

$$\Pr(\mathcal{L}(\mathsf{C}_n^\bullet)) = \Pr(\mathcal{L}(\mathsf{C}^\bullet) \mid \mathsf{C}^\bullet \in \mathcal{C}_n^\bullet).$$

By appaying Lemma 3.5 and Bayes's rule we obtain that there is a $C > 0$ such that

$$\Pr(\mathcal{L}(\mathsf{C}^\bullet) \mid \mathsf{C}^\bullet \in \mathcal{C}_n^\bullet) \leq C n^{\alpha-1} \Pr(\mathcal{L}(\mathsf{C}^\bullet), \mathsf{C}^\bullet \in \mathcal{C}_n^\bullet). \qquad (3.7)$$

By the last statement in Lemma 3.3 the event "$\mathcal{L}(\mathsf{C}^\bullet), \mathsf{C}^\bullet \in \mathcal{C}_n^\bullet$" implies, say, that $\max_{j \in [n]} Z_j \geq \frac{1}{2} \log n$ whenever n is sufficiently large. Since the Z_j's are Poisson distributed with parameter $\lambda_C > 0$ we obtain with Stirling's formula

$$\Pr\left(Z_j \geq \frac{1}{2} \log n \right) \leq \left(\frac{O(1)}{\log n} \right)^{\frac{1}{2} \log n} = n^{-\omega(1)}.$$

The statement then follows with room to spare from (3.7) by Markov's inequality. $\qquad\square$

Note that in the previous proof we were quite generous with the bound of $\log n$ – much more accurate estimates are possible here, but the derived bound is simple and will suffice for our purposes.

3.5 Simple vs complex graph classes

In this section we shall demonstrate that there are dramatic differences in the distribution of blocks in random graphs with n vertices from block-stable analytic classes of connected graphs. In particular, depending only on a single numerical parameter associated to the class we will see that the largest block in C_n has with high probability either $O(\log n)$ or a *linear* number $\Theta(n)$ of vertices.

Our starting point is the following observation. Let C be a block-stable analytic class of connected graphs. From Lemma 3.7 we infer that with high probability for any $0 < \varepsilon < 1$,

$$|b(\ell; C_n^{\bullet}) - b_{\ell}n| \leq \varepsilon b_{\ell}n \quad \text{for all } \ell \text{ with } b_{\ell}n = \omega(\log n) \text{ or } b_{\ell} = 0, \quad (3.8)$$

where $b_{\ell} = |\mathcal{B}'_{\ell-1}|y_C^{\ell-1}/(\ell-1)!$. As an auxiliary observation note that

$$\sum_{\ell \geq 2}(\ell-1)b_{\ell} = \sum_{\ell \geq 2}\frac{|\mathcal{B}'_{\ell-1}|y_C^{\ell-1}}{(\ell-2)!} = y_C B''(y_C). \quad (3.9)$$

For a given graph G with n vertices let $B(G)$ be set of blocks in G. Then from the block decomposition and by rooting G arbitrarily it follows that $n - 1 = \sum_{B \in B(G)}(|B| - 1)$. Let

$$S_n = \{2 \leq \ell \leq n : b_{\ell}n \geq \log^2 n \text{ or } b_{\ell}n = 0\}.$$

We obtain that with high probability

$$n - 1 = n\sum_{\ell \in S_n}(\ell-1)b_{\ell} + o(n) + H_n, \text{where } H_n = \sum_{\ell \in [n]\setminus S_n}(\ell-1)b(\ell; C_n^{\bullet}).$$

Since $|\mathcal{B}'_{\ell-1}| = |\mathcal{B}_{\ell}| \sim b\ell^{-\beta}\rho_B^{-\ell}\ell!$ we obtain for all $\ell \in \{2,\ldots,n\} \setminus S_n$ the estimate $b_{\ell} \sim \frac{b}{y_C}\ell^{-\beta+1}(y_C/\rho_B)^{\ell}$. Moreover, $y_C \leq \rho_B$, as by Lemma 3.2 $B'(y_C)$ is finite. Then, if the inequality is strict we clearly have $\ell b_{\ell} = o(1)$ for any $\ell \in \{2,\ldots,n\} \setminus S_n$. On the other hand, if $y_C = \rho_B$, then $b_{\ell} = O(\ell^{-\beta+1})$ and $\beta > 2$, as otherwise $B'(y_C)$ would not be finite; this implies in particular that also in this case $\ell b_{\ell} = o(1)$ for all $\ell \in \{2,\ldots,n\} \setminus S_n$. Writing $B''_{[n]}$ for the series containing the first n terms of B'' we thus get from the previous equation that with high probability

$$n - 1 = ny_C B''_{[n]}(y_C) + o(n) + \sum_{\ell \in [n]\setminus S_n}(\ell-1)b_{\ell} + H_n$$

$$= ny_C B''_{[n]}(y_C) + o(n) + H_n. \quad (3.10)$$

Since $H_n \geq 0$ and the previous relation is true for every n we infer (the deterministic statement) that

$$y_C B''(y_C) \leq 1.$$

What does this mean for our random graph $C_n^\bullet$? Actually, if $y_C B''(y_C) = 1$ then we immediately get that with high probability $H_n = o(n)$, that is, the number of vertices in "large" blocks (that is, those with size ℓ satisfying $b_\ell n \leq \log^2 n$) is $o(n)$. However, if $y_C B''(y_C) < 1$ then we get that $H_n \sim (1 - y_C B''(y_C))n$ with high probability, that is, there is a linear number of vertices in "large" blocks. We will actually see that in that case there is actually only *one* large block.

Theorem 3.9. *Let C be an analytic block-stable class of connected graphs and let $\xi = y_C B''(y_C)$. Then the following statements are true with high probability.*

1. *If $\xi < 1$, then the largest block in C_n contains $\sim (1 - \xi)n$ vertices. Moreover, $\beta > 3$ in this case, and each other block contains at most $O(n^{1/(\beta-2)})$ many vertices.*
2. *If $\xi = 1$ and further $y_C < \rho_B$, then the largest block in C_n contains at most $(\alpha + 1)\log_{\rho_B/y_C} n$ vertices.*
3. *In all other cases (i.e. $\xi = 1$ and $y_C = \rho_B$), the largest block contains $O(n^{1/(\beta-2)})$ vertices.*

Note that we obtain a striking dichotomy in the behavior of the largest block, which is reminiscent of the behavior of the largest connected component in the binomial random graph. Moreover, the third case in the previous theorem is stated just for completeness – at this point we have no example of a (natural) class satisfying the conditions.

Theorem 3.9 suggests the following general classification. If the class at hand belongs to the second category, that is, $\xi = 1$ and $y_C < \rho_B$, then a "typical" sample from C_n is a graph whose blocks have at most logarithmically many vertices. Observe that in this case, almost all *pairs* of vertices lie in different blocks, while this is not the case for graphs belonging to the first category in Theorem 3.9. A consequence of these facts is the following important observation. Random graphs from classes that belong to the second case "contain," in a well-defined sense, plenty of independence. In particular, any such graph can be generated by choosing independently every one of its blocks, and gluing them together at the cut-vertices. As the blocks contain only few vertices, and as they intersect each other only at single vertices, the impact of each block to the whole graph is small. Such random graphs resemble in a certain way the behavior of classical random graphs, where each edge is included independently, with the difference that here we choose the blocks independently of each other. However, random graphs from classes that belong to the first category do not have this property: a lot of structure that we cannot control is "hidden" in the giant block, which contains a constant fraction of the vertices. We say that graphs belonging to the second category have a *simple* structure, meaning that they look like a tree that is decorated with small blocks.

On the other hand, classes that belong to the first category are more *complex*, because a significant part of the graph sits inside the largest block. We will show in Section 3.6 one example of how in the *simple* case the particular structure can be exploited in order to study several other parameters, namely the degree distribution.

Before we give the proof of the theorem let us first discuss a few implications of it. We write $\text{Ex}(G_1, G_2, \ldots)$ for the class of connected graphs without $G_1, G_2, \ldots$ as minors. The work of Drmota et al. [17] implies that the classes $\text{Ex}(K_4, K_{2,3})$ (= outerplanar graphs), $\text{Ex}(K_4)$ (= series-parallel graphs) and $\text{Ex}(K_{2,3})$ are nice with $\alpha = \beta = 5/2$ and $\xi = 1$ and $y_C < \rho_B$, so that case 2 of Theorem 3.9 applies. In particular, a random graph from any of those classes has with high probability blocks of at most logarithmic size. Moreover, by using the main result in Giménez et al. [16], we infer that the class $\text{Ex}(K_5 \setminus e)$ has the same properties, and we are again in the second case. On the other hand, the classes $\text{Ex}(K_5, K_{3,3})$ (= planar graphs) and $\text{Ex}(K_{3,3})$ are nice with $\alpha = \beta = 7/2$ and $\xi < 1$, which follows from Giménez and Noy [13] and a result in Gerke et al. [25]. In particular, a uniform random (connected) planar graph contains a giant block with roughly cn vertices, where $c = 0.959\ldots$ is analytically given.

Proof (Theorem 3.9). We first treat the second case, that is, we assume that $\xi = y_C B''(y_C) = 1$ and $\rho_B > y_C$. Let $\mathcal{E}$ be the event that the largest block in $C_n^\bullet$ contains at most $s_n := (\alpha + 1) \log_{\rho_B / y_C} n$ many vertices. Then, by exchanging $C_n^\bullet$ with $C^\bullet$ we obtain that

$$\Pr(C_n^\bullet \notin \mathcal{E}) = \Pr(C^\bullet \notin \mathcal{E} \mid C^\bullet \in C_n^\bullet)$$

and further, by applying Bayes's rule and using Lemma 3.5, that there is a $C > 0$ such that

$$\Pr(C_n^\bullet \notin \mathcal{E}) \leq Cn^{\alpha - 1} \Pr(C^\bullet \notin \mathcal{E}, C^\bullet \in C_n^\bullet). \tag{3.11}$$

Using the last statement in Lemma 3.3 we obtain that the event "$C^\bullet \notin \mathcal{E}, C^\bullet \in C_n^\bullet$" implies that

$$\sum_{j=1}^{m} \mathbf{1}[|B_j'| \geq s_n - 1] \geq 1, \quad \text{where} \quad m = \sum_{j=1}^{n} K_j,$$

where the K_j's are independent and identically distributed $\text{Po}(\lambda_C)$ random variables and the B_j''s are independent random graphs following the Boltzmann distribution from $\mathcal{B}'$ with parameter y_C. The Chernoff bounds guarantee with room to spare that with probability at least $1 - o(n^{-\alpha})$ we have that $m \leq 2\lambda_C n$. Moreover, we will argue that

$$\mathbb{E}[\mathbf{1}[|B_j'| \geq s_n - 1]] = o(n^{-\alpha}). \tag{3.12}$$

From this, we get that

$$\Pr(C_n^{\bullet} \notin \mathcal{E}) \le C n^{\alpha-1} \Pr\left(\sum_{1 \le j \le 2\lambda_C n} \mathbb{1}[|B_j'| \ge s_n - 1] \ge 1 \right) + o(1).$$

By Markov's inequality and (3.12) the probability in the right-hand side is $o(n^{-\alpha+1})$, and the second statement in the theorem follows. To see (3.12), note that from the definition of the Boltzmann model and since C is nice we obtain that

$$\Pr(|B_j'| = \ell) = \frac{|\mathcal{B}_\ell'| y_C^\ell}{\ell! \lambda_C} = O(\ell^{-\beta+1} (y_C/\rho_B)^\ell). \tag{3.13}$$

For $\ell = s_n - 1$ this is $o(n^{-\alpha})$, and summing up the geometric series yields (3.12), as claimed. Moreover, note that the same argument as above yields the explicit bound

$$\Pr(\text{largest block in } C_n^{\bullet} \text{ has size} \le (\alpha+4) \log_{\rho_B/y_C} n) \ge 1 - O(n^{-3}), \tag{3.14}$$

a fact that we shall use later.

The proof of the third statement, that is, when $\xi = 1$ and $\rho_B = y_C$, is similar. There we obtain in complete analogy to (3.13) that

$$\Pr(|B_j'| = \ell) = \frac{|\mathcal{B}_\ell'| y_C^\ell}{\ell! \lambda_C} = O(\ell^{-\beta+1}),$$

and consequently $\Pr(|B_j'| \ge \ell) = O(\ell^{-\beta+2})$. Moreover, since $B''(\rho_B) = B''(y_C)$ is finite we infer that $\beta > 3$, and thus, for $\ell = \Omega(n^{1/(\beta-2)})$ we have that this probability is $O(1/n)$. The proof then is completed by a more careful estimate of the left-hand side in (3.11), where the factor $n^{\alpha-1}$ is replaced by $O(1)$; see Panagiotou et al. [20] for the details, and Janson [10], Equation (19.20), for a similar situation.

The proof of the first statement is more involved. We present here the main idea, and refer to Panagiotou and Steger [14] for all details. The first step is to obtain, as in (3.10), a statement about the total number S_n of vertices in "small" blocks, namely blocks of size at most $\alpha(n) n^{1/(\beta-2)}$, where $\alpha(n)$ is an arbitrary diverging function. The net result is that with high probability

$$S_n = y_C B''(y_C) n + o(n). \tag{3.15}$$

The matters complicating the proof of this fact is that actually, whenever $\ell = \omega(n^{1/(\beta-1)})$ we have that

$$\Pr(|B_j'| = \ell) = O(\ell^{-\beta+1}) = o(1/n),$$

and an application of Markov's inequality suggests for *any given* ℓ that there are no such blocks with high probability. However, note that a similar calculation yields

$$\Pr(|B'_j| \in \{\ell, \ell+1, \dots, 2\ell\}) = O(\ell^{-\beta+2}),$$

and this becomes $o(1/n)$ when $\ell = \omega(n^{1/(\beta-2)})$. An argument as in Lemma 3.7 then shows that the number of blocks with vertices in an any interval $\{\ell, \dots, 2\ell\}$ is close to $\sum_{\ell \le s \le 2\ell} b_s n$ with high probability for all ℓ in the desired range.

The second step is then to show that there is a single block with more than $\alpha(n)n^{1/(\beta-2)}$ vertices. Here we proceed with a combinatorial argument, where we compare directly the number of graphs in $\mathcal{C}_n^\bullet$ having exactly one block of size $N \sim (1 - y_\mathcal{C} B''(y_\mathcal{C}))n$, with the number of graphs having two or more blocks with more than $\alpha(n)n^{1/(\beta-2)}$ vertices. We shall not give the full argument here, and we refer to Panagiotou and Steger [14]; however, just to give the rough idea let us compare the number of graphs in $\mathcal{C}_n^\bullet$ with one block of size N and two blocks of size $N/2$ (assuming that N is even). Since the class is nice, the number of ways to choose a block with N vertices is $\sim bN^{-\beta}\rho_B^{-N}N!$, and the number of ways to choose the two blocks is

$$A := |\mathcal{B}_{N/2}|^2 \binom{N}{N/2} = O(N^{-2\beta}\rho_B^{-N}N!).$$

The binomial coefficient comes from the fact that the N labels have to be distributed among the two blocks. However, the two blocks may share a vertex, in which case the quantity A has to be multiplied by a factor of $O(N)$. Moreover, considering the block-tree of any graph in $\mathcal{C}_n^\bullet$, there are at most n additional ways as to where one of the blocks is rooted. In total, we get that we have to compare the quantites

$$bN^{-\beta}\rho_B^{-N}N! \quad \text{with} \quad O(N^{-2\beta+2}\rho_B^{-N}N!)$$

and the former is asymptotically larger than the latter whenever $-2\beta+2 < -\beta$, that is when $\beta > 2$. But we know that this is the case since $B'(\rho_B) < \infty$ by Lemma 3.2. The remaining cases – sizes sum up to N respecting the lower bound $\alpha(n)n^{1/(\beta-2)}$ and an arbitrary number ≥ 2 of blocks – can be treated systematically, and the statement follows. $\qquad\square$

3.6 Degree distribution for simple classes

Let $\mathcal{G}$ be a class of graphs. For any $G \in \mathcal{G}$ and any vertex v in G we write $d_G(v)$ for the degree of v in G. Moreover, for $k \in \mathbb{N}$ we set

$$D_k(G) = |\{v \text{ vertex of } T : d_G(v) = k\}|.$$

In this section we shall study the tails of the distribution of D_k if the graph is drawn uniformly at random from $\mathcal{C}_n$, where $\mathcal{C}$ is a simple block-stable class of connected graphs. Note that in complete analogy to the case of trees it does not make a difference if we consider rooted graphs with n vertices, as for every graph in C_n there are exactly n distinct graphs in $C_n^\bullet$ and moreover, rooting does not affect the value of D_k. We will prove the following statement, which in particular applies to outerplanar, series-parallel and $\mathrm{Ex}(K_5 \setminus e)$ graphs, see the discussion in Section 3.5.

Theorem 3.10. *Let C be a nice block-stable class of connected graphs such that $y_C B''(y_C) = 1$ and $y_C < \rho_B$. Then there is a constant $C > 0$ such that the following is true for all $n, k \in \mathbb{N}$. Let $\mu_{k,n} = \mathbb{E}[D_k(C_n)]$. Then for any $\varepsilon > 0$*

$$\Pr\left(|D_k(C_n) - \mu_{k,n}| \geq \varepsilon \mu_{k,n}\right) \leq C \frac{\log^2 n}{\varepsilon^2 \mu_{k,n}} + O(n^{-2}).$$

In other words, as long as $\mu_{k,n} = \omega(\log^2 n)$ we obtain a law of large numbers for $D_k(C_n)$ that is *universal* for the family of simple block-stable classes of connected graphs. In the proof we will make heavy use of the fact that in this case the largest block in a random graph contains only $O(\log n)$ vertices, see Theorem 3.9, which will allow us to derive a uniform bound of $O(\mu_{k,n} \log^2 n)$ for the variance of $D_k(C_n)$; from this the statement will follow by Chebyshev's inequality.

Proof of Theorem 3.10. We begin by defining an equivalence relation on the graphs in C_n. For a graph G we write $V(G)$ for its vertex set and we let $B(G)$ be the set of blocks in G. We say that two graphs G, G' with the same vertex set are *block-similar* and write $G \sim G'$, if the blocks in $B(G)$ and $B(G')$ contain the same vertices, that is, there is a bijection $\psi : B(G) \to B(G')$ such that for all $B \in B(G)$ we have that $V(B) = V(\psi(B))$. Note that if $G \sim G'$, then G and G' have the same number of blocks and the same set of cut-vertices; in particular the block-trees of G and G' are isomorphic graphs. Moreover, when G and G' are different graphs, then the edges in at least one block are distributed differently.

Note that $\sim$ is obviously an equivalence relation on a given set of graphs on the same vertex set, since bijective mappings are closed under inversion and composition. For C_n denote by $\mathcal{E}_n$ the equivalence classes under $\sim$, and for any graph $C \in C_n$ denote by $[C]$ its equivalence class. Moreover, for a graph G let $\mathrm{lb}(G)$ be the number of vertices in the largest block in G. Recall that $c(v; G)$ denotes the number of blocks in G that contain v, and let $\mathrm{lc}(G)$ be the largest such number. Set

$$\mathcal{S}_n = \{E \in \mathcal{E}_n : \mathrm{lb}(C) \leq (\alpha + 4) \log_{y_C/\rho_B} n$$

$$\text{and } \mathrm{lc}(C) \leq \log n \text{ for all } C \in E\}.$$

Since both the largest size of a block and the cut-degree of a vertex are invariant under rooting a graph we get from (3.14) and Lemma 3.8 that

$$\Pr([C_n] \in \mathcal{S}_n) \geq 1 - O(n^{-3}).$$

Let us write $\mathcal{B}_{n,k,\varepsilon}$ for the event $|D_k(C_n) - \mu_{k,n}| \geq \varepsilon \mu_{k,n}$. Since the equivalence classes are disjoint we then obtain that

$$\Pr(\mathcal{B}_{n,k,\varepsilon}) = \Pr(\mathcal{B}_{n,k,\varepsilon} \mid [C_n] \in \mathcal{S}_n) + O(n^{-3})$$
$$= \sum_{S \in \mathcal{S}_n} \Pr(\mathcal{B}_{n,k,\varepsilon} \mid [C_n] = S) \Pr([C_n] = S \mid [C_n] \in \mathcal{S}_n) + O(n^{-3}). \quad (3.16)$$

In order to study this probability we consider the random graph C_n under the condition that $[C_n] = S$, which we denote by C_S. Since the class $\mathcal{C}_n$ under consideration is block-stable, it is closed under the operation that takes an arbitrary block $B \in B(C)$ in any graph $C \in \mathcal{C}_n$ and exchanges it with some other block on the same vertex set $V(B)$ that is isomorphic to some block in $\mathcal{B}$. Moreover, applying this operation to a graph doesn't affect the equivalence class it belongs to, and all (distinct) graphs generated in this way are equiprobable. Thus, we may construct C_S as follows. Suppose that the graphs in S have M blocks, and that the set of labels of the ith block (where the numbering is arbitrary but fixed), $1 \leq i \leq M$, is denoted by V_i, then the distribution of C_S is the same as for the following experiment: for each $1 \leq i \leq M$ let $\mathsf{B}^{(i)}$ be an independent and uniform random graph from $\mathcal{B}[V_i]$ (here the species notation comes quite handy!); identify the induced subgraph of C_n on V_i with $\mathsf{B}^{(i)}$.

Let us write $\mu_S = \mathbb{E}[D_k(C_S)]$ and $V_S = \mathrm{Var}[D_k(C_S)]$. We will argue that there is a $C > 0$ such that $V_S \leq C \mu_S \log n$. For $v \in [n]$ let I_v be the indicator random variable that vertex v has degree k in C_S. Then

$$V_S = \sum_{u,v \in [n]} \left(\mathbb{E}[I_u I_v] - \mathbb{E}[I_u]\mathbb{E}[I_v] \right).$$

Note that whenever there is no $1 \leq i \leq M$ such that $u, v \in V_i$ then the corresponding summand equals zero by construction of C_S. Thus, let us denote for $v \in [n]$ by $N_S(v)$ the vertices in $[n]$ that are in a common block with v, that is

$$N_S(v) = \{u \in [n] : \text{there is an } 1 \leq i \leq M \text{ such that } u, v \in V_i\}.$$

Note that the definition of $\mathcal{S}_n$ guarantees that $N_S(v) = O(\log^2 n)$, since every vertex is contained in $O(\log n)$ blocks, and each block contains $O(\log n)$ vertices. Then

$$V_S \leq \sum_{v \in [n]} \sum_{u \in N_S(v)} \mathbb{E}[I_u I_v] \leq \mu_S + O(\log^2 n) \sum_{v \in [n]} \mathbb{E}[I_v] \leq C \mu_S \log^2 n$$

for an appropriate choice of $C > 0$. In total we obtain with Chebyshev's inequality that

$$\Pr(|D_k(C_n) - \mu_S| \geq \varepsilon \mu_S \mid [C_n] = S) \leq C \frac{\log^2 n}{\varepsilon^2 \mu_S}.$$

This is not quite what we would need in (3.16), since there we want to consider deviations from $\mu_{k,n}$ rather than from μ_S. For this, define the sets

$$S_n^{(c)} = \{S \in S_n : |\mu_{k,n} - \mu_S| \leq \varepsilon \mu_S/2\} \text{ and } S_n^{(f)} = S_n \setminus S_n^{(c)}$$

of equivalence classes that have an expected number of degree k vertices close to or far from $\mu_{k,n}$. Given any S in one of those classes, it is possible to compute μ_S: the definition of C_S guarantees that the probability that a vertex has degree k is a function of the sizes of the blocks in which it is contained. On the other hand, using the approach developed in Section 3.4 it is possible to derive accurate bounds for the number of vertices that are contained in a given number of blocks of certain sizes, which enables us to compute a "typical" value for the expected number of vertices with a given degree. In particular, the expectation, viewed as a random variable of the equivalence class, concentrates, implying that the contribution of the classes in $S_n^{(f)}$ to (3.16) vanishes. A precise analysis of the error bounds (which is omitted here) then yields the claimed statement. ☐

The expected number of vertices with a given degree in *simple* block-stable classes of graphs as well as much sharper tail bounds can be derived with the methods that were developed here, in particular with a finer study of the Boltzmann sampler for $C^\bullet$. This is done for example in the paper by Bernasconi et al. [18] where sharp estimates almost all the way up to the maximum degree are derived. For an alternative route treating only the case of bounded degrees but deriving a central limit theorem for D_k in this setting see Drmota et al. [17].

3.7 Further directions

As demonstrated in Section 3.6, the structure of random graphs from block-stable classes that typically consist only of small blocks can be exploited to study further parameters, as in this case most pairs of vertices are in disjoint blocks and thus independent in an appropriate probabilistic model. The same approach does not work for complex classes, since a significant fraction of the vertices is contained in the same block. This motivates a finer analysis regarding the typical structure of random 2-connected graphs; in particular, the central question is under which conditions 2-connected graphs can be decomposed into "building blocks" of higher complexity, in a way resembling the block-decomposition of connected graphs.

A recursive description of appropriate classes of 2-connected graphs goes back to Tutte [26] and consists of describing graphs in terms of building blocks of higher *connectivity*. For example, if we use the class of all 3-connected planar graphs as the base of the decomposition, then we obtain the classes consisting of all 2-connected planar graphs. Then a similar dichotomy as in Theorem 3.9 can be established, see Fountoulakis and Panagiotou [15] and Giménez et al. [16]: depending on some critical condition, which is given explicitly, it can be shown that a large random 2-connected graph either has only "small" 3-connected blocks, or a constant fraction of the vertices is contained in a single such building block. Hence, a picture that is completely analogous to the distribution of the block sizes in random connected graphs emerges.

All these results suggest the following classification scheme for a (block-stable) class $\mathcal{C}$ of connected graphs. If $\mathcal{C}$ is simple with respect to the block-decomposition then we are done. On the other hand, if it is complex then we may further consider the decomposition of the largest block in 3-connected graphs: the latter can also be classified into two further categories and we shall be saying that the class is *complex–simple* or *complex–complex*. Random graphs from complex–simple classes are amenable to probabilistic analysis, similar to what we performed in Section 3.6, but on the level of 3-connected components. On the other hand, a systematic treatment of random graphs from complex–complex classes (as for example the class of planar graphs) remains an important research challenge.

References

[1] Bollobás, Béla, *Random Graphs*, 2nd edn, Cambridge University Press, Cambridge, 2001.

[2] Janson, Svante, Łuczak, Tomasz, and Ruciński, Andrzej, *Random Graphs*, Wiley, New York, NY, 2000.

[3] Flajolet, Philippe and Sedgewick, Robert, *Analytic Combinatorics*. Cambridge University Press, Cambridge, 2009.

[4] Duchon, Philippe, Flajolet, Philippe, Louchard, Guy, and Schaeffer, Gilles, Boltzmann samplers for the random generation of combinatorial structures, *Comb. Probab. Comput.*, **13**(4–5) (2004), 577–625.

[5] Flajolet, Philippe, Fusy, Éric, and Pivoteau, Carine, Boltzmann sampling of unlabelled structures. Pages 201–211 of: *Proceedings of the Ninth Workshop on Algorithm Engineering and Experiments and the Fourth Workshop on Analytic Algorithmics and Combinatorics*. SIAM, Philadelphia, PA, 2007.

[6] Bodirsky, Manuel, Fusy, Éric, Kang, Mihyun, and Vigerske, Stefan, Boltzmann samplers, Pólya theory, and cycle pointing, *SIAM J. Comput.*, **40**(3) (2011), 721–769.

[7] Alon, Noga and Spencer, Joel H., *The Probabilistic Method. With an Appendix on the Life and Work of Paul Erdős*, 3rd edn, John Wiley & Sons, Hoboken, NJ, 2008.

[8] Davis, Burgess and McDonald, David, An elementary proof of the local central limit theorem, *J. Theor. Probab.*, **8**(3) (1995), 693–701.

[9] Drmota, Michael, *Random Trees. An Interplay between Combinatorics and Probability*, Springer, Wien, 2009.

[10] Janson, Svante, Simply generated trees, conditioned Galton-Watson trees, random allocations and condensation, *Probab. Surv.*, **9** (2012), 103–252.

[11] Takacs, Lajos, A generalization of the ballot problem and its application in the theory of queues, *J. Am. Stat. Assoc.*, **57** (1962), 327–337.

[12] Diestel, Reinhard, *Graph Theory. Graduate Texts in Mathematics*, vol. 173, Springer, 2010.

[13] Giménez, Omer and Noy, Marc, Asymptotic enumeration and limit laws of planar graphs, *J. Am. Math. Soc.*, **22**(2) (2009), 309–329.

[14] Panagiotou, Konstantinos and Steger, Angelika, Maximal biconnected subgraphs of random planar graphs, *ACM Trans. Algorithms*, **6**(2) (2010), 21.

[15] Fountoulakis, Nikolaos and Panagiotou, Konstantinos, 3-connected cores in random planar graphs, *Comb. Probab. Comput.*, **20**(3) (2011), 381–412.

[16] Giménez, Omer, Noy, Marc, and Rué, Juanjo, Graph classes with given 3-connected components: asymptotic enumeration and random graphs, *Random Struct. Algorithms*, **42**(4) (2013), 438–479.

[17] Drmota, Michael, Fusy, Éric, Kang, Mihyun, Kraus, Veronika, and Rué, Juanjo, Asymptotic study of subcritical graph classes, *SIAM J. Discrete Math.*, **25**(4) (2011), 1615–1651.

[18] Bernasconi, Nicla, Panagiotou, Konstantinos, and Steger, Angelika, The degree sequence of random graphs from subcritical classes, *Comb. Probab. Comput.*, **18**(5) (2009), 647–681.

[19] Drmota, Michael, Giménez, Omer, and Noy, Marc, Degree distribution in random planar graphs, *J. Comb. Theory, Ser. A*, **118**(7) (2011), 2102–2130.

[20] Panagiotou, Konstantinos, Stufler, Benedikt, and Weller, Kerstin, Scaling limits of random graphs from subcritical classes, *Ann. of Prob.* Accepted for publication.

[21] Joyal, André, Une théorie combinatoire des séries formelles, *Adv. in Math.*, **42**(1) (1981), 1–82.

[22] Bergeron, F., G. Labelle, and P. Leroux, *Combinatorial Species and Tree-like Structures. Encyclopedia of Mathematics and Its Applications*, vol. 67. Cambridge University Press, Cambridge, 1998. Translated from the 1994 French original by Margaret Readdy, with a foreword by Gian-Carlo Rota.

[23] Fusy, Éric, Uniform random sampling of planar graphs in linear time, *Random Struct. Algorithms*, **35**(4) (2009), 464–522.

[24] Harary, Frank and Palmer, Edgar M., *Graphical Enumeration*. Academic Press, New York-London, 1973.

[25] Gerke, Stefanie, Giménez, Omer, Noy, Marc, and Weißl, Andreas, The number of graphs not containing $K_{3,3}$ as a minor, *Electron. J. Comb.*, **15**(1) (2008), research paper r114, 20.

[26] Tutte, William Thomas, *Connectivity in Graphs*. University of Toronto Press, Toronto, (1996).

3

Lectures on random geometric graphs

Mathew Penrose

These notes are based on a series of five lectures given at the LMS-EPSRC Short Course in Birmingham, August 2013. The notes complement the author's monograph on this topic [1]; there is an overlap of material but here it is presented in a selective and often simplified manner, making these notes suitable for an introductory course of study on the subject. Material covered includes the limit theory (Poisson and normal) for the number of edges, the maximum degree, connectivity and Hamiltonian paths; the last of these topics (Section 6) was not covered in [1], or in the 2013 lectures.

These notes are pitched at the level of a first-year graduate mini-course. They should be accessible to a reader with knowledge of probability theory at the level of a final-year undergraduate or an entry-level graduate course, along with a few basic definitions in graph theory. Each section of these notes contains a few exercises at the end. Solutions are provided in the final section.

1 Introduction

There has been great interest recently in mathematical modeling of large networks, often via random graphs. However, many of the most intensively studied random graph models have no spatial content. Since many real-world networks have nodes with spatial locations which influence the graph structure, it is of interest to seek to model such networks in a manner which takes this into account.

The following provides a simple model for networks with spatial content; vertices have a spatial location and only nearby vertices are connected together.

Given finite $\mathcal{X} \subset \mathbb{R}^d$, and $r > 0$, the *geometric graph* $G(\mathcal{X}, r)$ has vertex set $\mathcal{X}$ and edge set $\{\{x, y\} : \|x - y\| \leq r\}$, where $\| \cdot \|$ is the Euclidean norm.

For example, the vertices might represent wireless transmitters with range r. Or they might represent trees in an orchard in which a disease may pass between trees up to range r. The idea of making sense of a point pattern by connecting nearby points with edges is an ancient one; consider the constellations of stars.

A *random* geometric graph (also known as a *Gilbert graph*) is obtained by taking $\mathcal{X}$ to be a random set of points.

Let $\xi_1, \xi_2, \ldots$ be independent random d-vectors, uniformly distributed over $[0, 1]^d$ (typically $d = 2$). Set

$$\mathcal{X}_n := \{\xi_1, \ldots, \xi_n\}$$

and

$$\mathcal{P}_n := \{\xi_1, \ldots, \xi_{N_n}\} \qquad (1.1)$$

with N_n Poisson distributed with parameter n, independent of $(\xi_1, \xi_2, \ldots)$. Then (see Exercise 1.1) $\mathcal{P}_n$ is a *Poisson point process* in $[0, 1]^d$ with n times Lebesgue measure as its mean measure, i.e. for Borel $A, A_1, \ldots, A_n \subset [0, 1]^d$

$$\mathcal{P}_n(A) \sim \mathrm{Po}[n|A|]; \qquad (1.2)$$

$$\mathcal{P}_n(A_1), \ldots, \mathcal{P}_n(A_k) \text{ are independent for } A_1, \ldots, A_k \text{ disjoint,} \qquad (1.3)$$

where $\mathcal{X}(A)$ (for any point set $\mathcal{X}$ and any $A \subset \mathbb{R}^d$) means the number of points of $\mathcal{X}$ in A, and $|\cdot|$ here denotes Lebesgue measure, and for $\lambda > 0$, $\mathrm{Po}(\lambda)$ denotes the Poisson distribution with parameter λ.

The random geometric graphs we consider in these lecture notes are $G(\mathcal{X}_n, r_n)$ and $G(\mathcal{P}_n, r_n)$, with $(r_n)_{n \geq 1}$ a specified sequence of distance parameters. Random geometric graphs over nonuniformly distributed points have also been considered, for example in [1], but in these notes we consider only the case of uniformly distributed points.

One reason to study random geometric graphs is to explore "typical" properties of geometric graphs. Another reason is to assess statistical tests based on the graph $G(\mathcal{X}_n, r_n)$, for example tests for uniformity. It is of interest to compare this random graph model with others, such as the Erdős–Rényi random graph $G(n, p)$.

Notation. Many of the results described in this course are asymptotic results as $n \to \infty$. Unless stated otherwise, any limiting statement in the sequel is as $n \to \infty$. Also, for positive real-valued sequences a_n and b_n we use the following asymptotic notational conventions:

- $a_n = O(b_n)$ means $\limsup(a_n/b_n) < \infty$.
- $a_n = \Theta(b_n)$ means that both $a_n = O(b_n)$ and $b_n = O(a_n)$.
- $a_n = o(b_n)$ means that $a_n/b_n \to 0$. This may also be written as $a_n \ll b_n$ or as $b_n \gg a_n$, or as $b_n = \omega(a_n)$.
- $a_n \sim b_n$ means $a_n/b_n \to 1$.

Let θ denote the volume of the unit ball in $\mathbb{R}^d$.

Given two points $x, y \in \mathbb{R}^d$, we shall say x lies *to the left* of y if x precedes y in the lexicographic ordering on $\mathbb{R}^d$.

If $r_n \to 0$, the expected number of edges incident to a "typical vertex" of $\mathcal{X}_n$ or $\mathcal{P}_n$ is asymptotic to $\theta n r_n^d$ (in the case of $\mathcal{X}_n$, Exercise 1.2 makes this statement precise).

As this suggests, we often get different limiting behavior depending on the limit behavior of $n r_n^d$. We refer to cases with $n r_n^d \to 0$ as the *sparse limit*, $n r_n^d \to \infty$ as the *dense limit* and $n r_n^d = \Theta(1)$ as the *thermodynamic limit*.

Exercise 1.1. Prove (1.2) and (1.3).

Exercise 1.2. Let $D_{1,n}$ denote the degree of vertex ξ_1 in the graph $G(\mathcal{X}_n, r_n)$. Prove that if $r_n \to 0$, then $\mathbb{E}[D_{1,n}] \sim \theta n r_n^d$ as $n \to \infty$.

2 Edge counts

Let $\mathcal{E}_n$ denote the number of edges of $G(\mathcal{X}_n, r_n)$, and let $\mathcal{E}_n'$ be the number of edges of $G(\mathcal{P}_n, r_n)$. We consider the limiting behavior of the probability distribution of $\mathcal{E}_n$ and $\mathcal{E}_n'$. More generally one could (by similar methods) also consider the limiting distribution of the number of subgraphs isomorphic to some specified connected finite graph; see Penrose [1] for details.

Note that if $r_n \to 0$ then (see Exercise 2.1)

$$\mathbb{E}\mathcal{E}_n \sim \theta(n^2 r_n^d)/2 \tag{2.1}$$

and

$$\mathbb{E}\mathcal{E}_n' \sim \theta(n^2 r_n^d)/2. \tag{2.2}$$

First we consider the sparse limit ($n r_n^d \to 0$). In this case we can show that $\mathcal{E}_n$ is well approximated by a Poisson distributed variable (though if also $n^2 r_n^d$ is large this is itself well approximated by a normal random variable).

We prove this using the technique of *dependency graphs*. Suppose $(V, \sim)$ is a finite graph without loops, i.e. for $\alpha, \beta \in V$ we write $\alpha \sim \beta$ if α, β are adjacent. Since we assume there are no loops, we have $\alpha \not\sim \alpha$ for all vertices α.

This is a *dependency graph* for a set of random variables $(W_\alpha, \alpha \in V)$ if whenever $A \subset V, B \subset V$ with $A \cap B = \emptyset$ and no edges connecting A to B,

$$(W_\alpha, \alpha \in A) \text{ is independent of } (W_\beta, \beta \in B).$$

Lemma 2.1. (Poisson Approximation Lemma). *Suppose* $(V, \sim)$ *is a finite graph and* $(W_\alpha)_{\alpha \in V}$ *is a family of* $0 - 1$ *valued random variables, having* $(V, \sim)$ *as a dependency graph. For* $\alpha, \beta \in V$ *set* $p_\alpha = \mathbb{P}[W_\alpha = 1]$ *and* $p_{\alpha\beta} = \mathbb{P}[W_\alpha = 1, W_\beta = 1]$. *Then setting* $W = \sum_{\alpha \in V} W_\alpha$ *and* $\lambda = \sum_{\alpha \in V} p_\alpha$, *we have*

$$\sum_{k=0}^{\infty} |\mathbb{P}[W = k] - e^{-\lambda} \lambda^k / k!| \leq \min(2/\lambda, 6)$$

$$\times \left(\sum_{\alpha \in V} p_\alpha^2 + \sum_{\alpha} \sum_{\beta \sim \alpha} (p_{\alpha\beta} + p_\alpha p_\beta) \right). \tag{2.3}$$

Proof. Omitted - see [1, Theorem 2.1]. The result was originally due to Arratia, Goldstein and Gordon.

The next result is a special case of [1, Theorem 3.4]. We use $\overset{\mathcal{D}}{\longrightarrow}$ to denote convergence in distribution (as $n \to \infty$).

Theorem 2.2. *Suppose* $nr_n^d \to 0$. *Let* $\lambda_n = \mathbb{E}\mathcal{E}_n$. *Then*

$$\sum_{k=0}^{\infty} |\mathbb{P}[\mathcal{E}_n = k] - e^{-\lambda_n} \lambda_n^k / k!| = O(nr_n^d). \tag{2.4}$$

In particular, if $\lambda_n \to \alpha \in (0, \infty)$, *then* $\mathcal{E}_n \overset{\mathcal{D}}{\longrightarrow} \text{Po}(\alpha)$; *moreover, in this case* $\mathcal{E}_n' \overset{\mathcal{D}}{\longrightarrow} \text{Po}(\alpha)$.

Equation (2.4) provides an error bound in the Poisson approximation of $\mathcal{E}_n$. See Exercise 2.2 for a similar error bound in the case of $\mathcal{E}_n'$.

Proof of Theorem 2.2. Let $V = \{\alpha = \{i, j\} : 1 \leq i < j \leq n\}$ with $\alpha \sim \beta$ if $\alpha \cap \beta \neq \emptyset$ and $\alpha \neq \beta$. Set

$$W_{\{i,j\}} = \mathbf{1}\{\|\xi_i - \xi_j\| \leq r_n\}.$$

Then $\mathcal{E}_n = \sum_{\alpha \in V} W_\alpha$ and $(V, \sim)$ is a dependency graph for $\{W_\alpha\}$. Now p_α depends on n but not α and

$$p_\alpha \sim \theta r_n^d,$$

and similarly for $\alpha \sim \beta$ we have

$$p_{\alpha\beta} \sim (\theta r_n^d)^2,$$

so that

$$\lambda_n = \sum_{\alpha \in V} p_\alpha \sim \binom{n}{2} \theta r_n^d \sim \frac{n^2 \theta r_n^d}{2}$$

and

$$\sum_{\alpha \in V} p_\alpha^2 \sim \binom{n}{2} (\theta r_n^d)^2 \sim \lambda_n (\theta r_n^d),$$

while

$$\sum_{\alpha \in V} \sum_{\beta \sim \alpha} (p_{\alpha\beta} + p_\alpha p_\beta) \sim \binom{n}{2} \times 2(n-2) \times 2\theta^2 r_n^{2d}$$

$$= O(\lambda_n n r_n^d).$$

Thus Lemma 2.1 gives us (2.4).

Now suppose $\lambda_n \to \alpha \in (0, \infty)$. Then by (2.1) we have $n^2 r_n^d \to 2\alpha/\theta$, so that $n r_n^d \to 0$. Hence we have (2.4), and thus $\mathcal{E}_n \xrightarrow{\mathcal{D}} \mathrm{Po}(\alpha)$. It remains to show that $\mathcal{E}'_n \xrightarrow{\mathcal{D}} \mathrm{Po}(\alpha)$. Using (1.1), we have for any $j \in \mathbb{N}_0 := \mathbb{N} \cup \{0\}$ that

$$\mathbb{P}[\mathcal{E}'_n = j] - \frac{e^{-\alpha}\alpha^j}{j!} = \sum_{k=0}^{\infty} \mathbb{P}[N_n = k] \left(\mathbb{P}[\mathcal{E}_{k,n} = j] - \frac{e^{-\alpha}\alpha^j}{j!} \right),$$

where $\mathcal{E}_{k,n}$ denotes the number of edges of $G(\mathcal{X}_k, r_n)$. By Chebyshev's inequality, $\mathbb{P}[|N_n - n| \geq n^{3/4}] \to 0$, and therefore to show $\mathcal{E}'_n \xrightarrow{\mathcal{D}} \mathrm{Po}(\alpha)$ it suffices to show that for any $j \in \mathbb{N}_0$ we have

$$\lim_{n \to \infty} \sup_{n - n^{3/4} \leq k \leq n + n^{3/4}} \left| \mathbb{P}[\mathcal{E}_{k,n} = j] - \frac{e^{-\alpha}\alpha^j}{j!} \right| = 0. \qquad (2.5)$$

For any sequence $k(n)$ with $n - n^{3/4} \leq k(n) \leq n + n^{3/4}$ for all n, by (2.1), since $k(n) \sim n$ we have $\mathbb{E}\mathcal{E}_{k(n),n} \to \alpha$, so by the result already proved we have $\mathbb{P}[\mathcal{E}_{k(n),n} = j] \to \frac{e^{-\alpha}\alpha^j}{j!}$ as $n \to \infty$, and this gives us (2.5) as required. $\qquad \square$

The following lemma will frequently be useful. It is called "Palm theory for the Poisson process" in [1] but here we call it the *Mecke formula*. In the proof we use notation

$$(n)_k := n(n-1) \cdots (n-k+1) \quad \text{for } n, k \in \mathbb{N}$$

(the so-called *descending factorial*). Also, in the following formula (and elsewhere) the region of integration, when not specified otherwise, is to be taken to be $[0, 1]^d$.

Lemma 2.3. (Mecke formula) *Let $k \in \mathbb{N}$. For any measurable real-valued function f, defined on the product of $(\mathbb{R}^d)^k$ and the space of finite subsets of $[0,1]^d$, for which the following expectation exists,*

$$\mathbb{E} \sum_{X_1,\ldots,X_k \in \mathcal{P}_n}^{\neq} f(X_1, X_2, \ldots, X_k, \mathcal{P}_n \setminus \{X_1, \ldots, X_k\})$$

$$= n^k \int dx_1 \cdots \int dx_k \mathbb{E} f(x_1, \ldots, x_k, \mathcal{P}_n)$$

where $\sum^{\neq}$ means the sum is over ordered k-tuples of distinct points of $\mathcal{P}_n$.

Proof. We condition on the number of points of $\mathcal{P}_n$. Then

$$\mathbb{E} \sum_{X_1,\ldots,X_k \in \mathcal{P}_n}^{\neq} f(X_1, X_2, \ldots, X_k, \mathcal{P}_n \setminus \{X_1, \ldots, X_k\})$$

$$= \sum_{m=k}^{\infty} \left(e^{-n} \frac{n^m}{m!} \right) (m)_k \int dx_1 \cdots \int dx_m f(x_1, \ldots, x_k, \{x_{k+1}, \ldots, x_m\})$$

$$= n^k \int dx_1 \cdots \int dx_k \sum_{m=k}^{\infty} \left(\frac{e^{-n} n^{m-k}}{(m-k)!} \right)$$

$$\times \int dy_1 \cdots \int dy_{m-k} f(x_1, \ldots, x_k, \{y_1, \ldots, y_{m-k}\})$$

$$= n^k \int dx_1 \cdots \int dx_k \sum_{r=0}^{\infty} \left(\frac{e^{-n} n^r}{r!} \right)$$

$$\times \int dy_1 \cdots \int dy_r f(x_1, \ldots, x_k, \{y_1, \ldots, y_r\})$$

$$= n^k \int dx_1 \cdots \int dx_k \mathbb{E} f(x_1, \ldots, x_k, \mathcal{P}_n)$$

where first we have made the substitution $y_j = x_{k+j}$ for $k < j \leq m$, and then we have set $r = m - k$. $\square$

Exercise 2.1. Prove (2.1) and (2.2).

Exercise 2.2. Show that under the hypothesis of Theorem 2.2, setting $\lambda_n' := \mathbb{E}\mathcal{E}_n'$, we have

$$\sum_{k=0}^{\infty} |\mathbb{P}[\mathcal{E}_n' = k] - e^{-\lambda_n'} (\lambda_n')^k / k!| = O(nr_n^d). \qquad (2.6)$$

[Hint: Given n, discretize space into squares of side $1/m$ and let $m \to \infty$ with n fixed. Use Lemma 2.1 and the spatial independence of the Poisson process.]

3 Edge counts: normal approximation

Given $x \in \mathbb{R}^d$ and $r > 0$, let $B(x;r)$ denote the closed Euclidean ball of radius r centred at x.

We now give limiting behavior of the variance of $\mathcal{E}'_n$ in thermodynamic or dense limit. For a more general result (considering the number of induced subgraphs isomorphic to a specified connected graph, rather than just the number of edges), see [1, Proposition 3.7].

Proposition 3.1. *If* $\liminf(nr_n^d) > 0$ *and* $r_n \to 0$ *then*

$$\mathrm{Var}(\mathcal{E}'_n) \sim n[(n\theta r_n^d)^2 + (1/2)n\theta r_n^d].$$

Note that in the dense limit this simplifies to $\mathrm{Var}(\mathcal{E}'_n) \sim n(\theta n r_n^d)^2$. The proof of Proposition 3.1 uses the following fact (see Exercise 3.1).

Lemma 3.2. *If X is a Poisson variable with parameter λ, and $k \in \mathbb{N}$, then* $\mathbb{E}[(X)_k] = \lambda^k$.

Proof of Proposition 3.1. Let $g_n(x,y) = \mathbf{1}\{\|x - y\| \leq r_n\}$ for $x, y \in \mathbb{R}^d$. Then by the Mecke formula

$$2\mathbb{E}\mathcal{E}'_n = \mathbb{E} \sum_{X,Y \in \mathcal{P}_n}^{\neq} g_n(X,Y) = n^2 \int \int g_n(x,y) dx\, dy \tag{3.1}$$

with all integrals taken over $[0,1]^d$ in this proof. Thus,

$$\mathbb{E}\mathcal{E}'_n \sim n^2 \theta r_n^d / 2. \tag{3.2}$$

Now consider $(\mathcal{E}'_n)^2$. This is the number of ordered pairs of edges in $G(\mathcal{P}_n, r_n)$. We may decompose this as

$$(\mathcal{E}'_n)^2 = S_{n,0} + S_{n,1} + S_{n,2}$$

where for $i = 0, 1, 2$ we let $S_{n,i}$ denote the number of ordered pairs of edges with i endpoints in common. Then

$$S_{n,0} = \frac{1}{4} \sum_{U,V,X,Y \in \mathcal{P}_n}^{\neq} g_n(U,V) g_n(X,Y),$$

where $\sum^{\neq}$ means the sum is over ordered 4-tuples of distinct points in $\mathcal{P}_n$. By the Mecke formula, followed by (3.1),

$$\mathbb{E}S_{n,0} = \frac{n^4}{4} \int \int \int \int g_n(u,v) g_n(x,y) du\, dv\, dx\, dy = (\mathbb{E}\mathcal{E}'_n)^2.$$

Also, $S_{n,2} = \mathcal{E}'_n$ and therefore

$$\mathrm{Var}(\mathcal{E}'_n) = \mathbb{E}[(\mathcal{E}'_n)^2] - (\mathbb{E}\mathcal{E}'_n)^2 = \mathbb{E}S_{n,1} + \mathbb{E}\mathcal{E}'_n. \tag{3.3}$$

Next consider $S_{n,1}$. We have

$$S_{n,1} = \sum_{X \in \mathcal{P}_n} \sum_{Y,Z \in \mathcal{P}_n \setminus \{X\}}^{\neq} g_n(X,Y)g_n(X,Z) = \sum_{X \in \mathcal{P}_n} h_n(X, \mathcal{P}_n \setminus \{X\}),$$

where for finite $\mathcal{X} \subset \mathbb{R}^d$ we set $h_n(x, \mathcal{X}) := (\mathcal{X}(B(x; r_n)))_2$ and $(n)_2 := n(n-1)$ is the descending factorial. Using the Mecke formula, we have

$$\mathbb{E}S_{n,1} = n \int \mathbb{E}h_n(x, \mathcal{P}_n)dx.$$

Now using Lemma 3.2, and dominated convergence, we have

$$\mathbb{E}S_{n,1} = n \int (n|B(x; r_n) \cap [0,1]^d|)^2 dx \sim n(n\theta r_n^d)^2. \qquad (3.4)$$

By combining (3.3), (3.4) and (3.2) we get the result. $\qquad \square$

We shall prove a central limit theorem for $\mathcal{E}_n'$ in the thermodynamic or dense limit, using the following result of Chen and Shao [2] (alternatively one could use [1, Theorem 2.4], which would give a slower rate of convergence in the central limit theorem). We let Φ be the standard normal cumulative distribution function, i.e. $\Phi(x) = (2\pi)^{-1/2} \int_{-\infty}^{x} e^{-t^2/2}dt$ for $x \in \mathbb{R}$.

Lemma 3.3. (Normal Approximation Lemma; see [2, Theorem 2.7]) *Let $D \in \mathbb{N}$. Let W_i, $i \in V$, be random variables indexed by the vertices of a finite dependency graph with $|V|$ vertices, all of degree at most D. Let $W = \sum_{i \in V} W_i$. Assume that $\mathbb{E}[W^2] = 1, \mathbb{E}[W_i] = 0$, and for some $\beta > 0$, that $\mathbb{E}[|W_i|^3] \leq \beta$ for all $i \in V$. Then*

$$\sup_{t \in \mathbb{R}} |P[W \leq t] - \Phi(t)| \leq 75D^{10}|V|\beta.$$

Proof. Omitted. Lemmas 3.3 and 2.1 are both proved by versions of *Stein's method*, which is an important topic in its own right, but beyond the scope of these lecture notes.

Now we can give our central limit theorem.

Theorem 3.4. *Suppose $r_n \to 0$ but $\liminf(nr_n^d) > 0$. Then*

$$\sup_{x \in \mathbb{R}} \left| \mathbb{P}\left[\frac{\mathcal{E}_n' - \mathbb{E}\mathcal{E}_n'}{\sqrt{\mathrm{Var}(\mathcal{E}_n')}} \leq x \right] - \Phi(x) \right| = O(r_n^{d/2}) \quad as \ n \to \infty. \qquad (3.5)$$

In particular, $(\mathcal{E}_n' - \mathbb{E}\mathcal{E}_n')/\sqrt{\mathrm{Var}(\mathcal{E}_n')} \xrightarrow{\mathcal{D}} \mathcal{N}(0,1)$, where $\mathcal{N}(0,1)$ denotes a random variable with distribution function Φ.

Proof. Given n, partition $\mathbb{R}^d$ into cubes of side r_n, and let those cubes in the partition that have nonempty intersection with $[0,1]^d$ be denoted $C_1, \ldots, C_{k_n}$.

Then $k_n \sim r_n^{-d}$. Let $M_i := M_i(n)$ denote the number of edges of $G(\mathcal{P}_n, r_n)$ with left-endpoint in C_i. Set

$$W_i := W_i(n) := \frac{M_i - \mathbb{E}M_i}{\sqrt{\mathrm{Var}(\mathcal{E}'_n)}}.$$

Then

$$\frac{\mathcal{E}'_n - \mathbb{E}\mathcal{E}'_n}{\sqrt{\mathrm{Var}(\mathcal{E}'_n)}} = \sum_{i=1}^{k_n} W_i.$$

Set $V = \{1, 2, \ldots, k_n\}$ and for $i, j \in V$ put $i \sim j$ if C_i and C_j are neighboring cubes (i.e. they touch, so allowing diagonal neighbors) or they have a common neighbor. Then $(V, \sim)$ is a dependency graph for $(W_i, i \in V)$ and the maximal degree of this graph is at most $5^d - 1$, independent of n.

By Proposition 3.1 we have that $\mathrm{Var}(\mathcal{E}'_n) = \Theta(n(nr_n^d)^2)$. Therefore, by Lemma 3.3 it suffices to prove that

$$k_n \max_{1 \le i \le k_n} \frac{\mathbb{E}[|M_i - \mathbb{E}M_i|^3]}{(n^{1/2}nr_n^d)^3} = O(r_n^{d/2}). \tag{3.6}$$

We shall estimate $\mathbb{E}[|M_i - \mathbb{E}M_i|^3]$ by first estimating $\mathbb{E}[|M_i - \mathbb{E}M_i|^4]$ (which is the same as $\mathbb{E}[(M_i - \mathbb{E}M_i)^4]$), and then using Hölder's inequality. By the binomial theorem,

$$\mathbb{E}[(M_i - \mathbb{E}M_i)^4] = \sum_{j=0}^{4} \binom{4}{j} (-\mathbb{E}M_i)^{4-j} \mathbb{E}[M_i^j]. \tag{3.7}$$

By the Mecke equation,

$$\mathbb{E}M_i = n^2 \int \int g_{i,n}(x, y) dx \, dy, \tag{3.8}$$

where $g_{i,n}(x, y)$ is the indicator of the statement that $\|x - y\| \le r_n$ and also $x \in C_i$ and also x lies to the left of y (recall that all integrals are over $[0, 1]^d$ unless specified otherwise).

Note that

$$\mathbb{E}[M_i^4] = \mathbb{E} \sum_e \sum_{e'} \sum_{e''} \sum_{e'''} 1 \tag{3.9}$$

where each sum runs through all edges of $G(\mathcal{P}_n, r_n)$ having left-endpoint in the cube C_i. The leading-order term in this expectation comes from when all of e, e', e'', e''' have distinct endpoints, so that this leading-order term is

$$\mathbb{E} \sum_{X_1, Y_1, \ldots, X_4, Y_4 \in \mathcal{P}_n}^{\neq} g_{i,n}(X_1, Y_1) \cdots g_{i,n}(X_4, Y_4)$$

Therefore by the Mecke formula, the leading term equals

$$n^8 \int \cdots \int dx_1 \cdots dx_4 dy_1 \cdots dy_4 \prod_{i=1}^{4} g_{i,n}(x_i, y_i)$$

$$= \left(n^2 \int \int dx dy g_{i,n}(x, y) \right)^4$$

$$= (\mathbb{E}[M_i])^4,$$

where the last line comes from (3.8).

Similarly, the leading-order term in $\mathbb{E}M_i^3$ (coming from triples of edges with no endpoints in common) is equal to $(\mathbb{E}[M_i])^3$, and the leading-order term in $\mathbb{E}M_i^2$ (coming from pairs of edges with no endpoints in common) is equal to $(\mathbb{E}[M_i])^2$. Therefore if we collect together all the leading-order terms in (3.7) we get

$$\sum_{j=0}^{4} \binom{4}{j} (-\mathbb{E}M_i)^{4-j} (\mathbb{E}M_i)^j$$

which is equal to $(\mathbb{E}M_i - \mathbb{E}M_i)^4$ and therefore comes to zero.

Next we consider second-order terms. Let R_i be the number of ordered pairs of edges having a common endpoint and both with left endpoint in C_i. Then

$$R_i = \sum_{X,Y,Z \in \mathcal{P}_n}^{\neq} (g_{i,n}(X, Y) + g_{i,n}(Y, X))(g_{i,n}(X, Z) + g_{i,n}(Z, X))$$

and by the Mecke formula,

$$\mathbb{E}R_i = n^3 \int \int \int (g_{i,n}(x, y) + g_{i,n}(y, x))(g_{i,n}(x, z) + g_{i,n}(z, x))$$
$$\times dx\, dy\, dz. \tag{3.10}$$

The second-order term in the expression (3.9) for $\mathbb{E}[M_i^4]$ comes from quadruples of edges e, e', e'', e''' having seven distinct endpoints among them (so having precisely one common endpoint among the four edges). There are six ways to choose which two of the edges e, e', e'', e''' share an endpoint, and therefore the second-order term in $\mathbb{E}[M_i^4]$ comes to

$$6\mathbb{E} \sum_{X,Y,Z,X_1,X_2,Y_1,Y_2 \in \mathcal{P}_n}^{\neq} (g_{i,n}(X, Y) + g_{i,n}(Y, X))$$

$$\times (g_{i,n}(X, Z) + g_{i,n}(Z, X)) g_{i,n}(X_1, Y_1) g_{i,n}(X_2, Y_2)$$

and by the Mecke formula this comes to

$$6n^7 \int \cdots \int (g_{i,n}(x,y) + g_{i,n}(y,x))(g_{i,n}(x,z) + g_{i,n}(z,x))$$

$$\times g_{i,n}(x_1,y_1)g_{i,n}(x_2,y_2)dxdydzdx_1dy_1dx_2dy_2$$

$$= 6(\mathbb{E}M_i)^2\mathbb{E}R_i,$$

where for the last line we have used (3.8) and (3.10).

Similarly, the second-order term in $\mathbb{E}[M_i^3]$ comes to $3\mathbb{E}M_i\mathbb{E}R_i$ (see Exercise 3.3), and the second-order term in $\mathbb{E}M_i^2$ comes to $\mathbb{E}R_i$. There is no second-order term in $\mathbb{E}M_i$.

Thus the sum of all second-order terms in (3.7) comes to

$$(\mathbb{E}M_i)^2\mathbb{E}R_i\left(\binom{4}{4}\times 6 - \binom{4}{3}\times 3 + \binom{4}{2}\times 1\right) = 0.$$

Therefore, the leading-order nonzero term in (3.7) comes from the third-order terms. The third-order term in $\mathbb{E}M_i^4$ comes from 4-tuples of edges e, e', e'', e''' having two shared endpoints among them (so with a total of six distinct endpoints). This is bounded by some combinatorial constant times

$$\sum_{X_1,\ldots,X_6 \in \mathcal{P}_n}^{\neq} h_{i,n}^*(X_1,\ldots,X_6)$$

where $h_{i,n}^*(x_1,\ldots,x_6)$ is the indicator of the event that $x_1,\ldots,x_6$ all lie in C_i or in one of the neighboring cubes. Therefore by the Mecke formula, the third-order term in $\mathbb{E}M_i^4$ is bounded by a constant times

$$n^6 \int \cdots \int h_{i,n}^*(x_1,\ldots,x_6)dx_1\cdots dx_6$$

and therefore is $O((nr_n^d)^6)$. Similarly, the higher-order terms are $O((nr_n^d)^5)$. Combining all this, we find from (3.7) that

$$\mathbb{E}[(M_i - \mathbb{E}M_i)^4] = O((nr_n^d)^6),$$

and also by Hölder's inequality for any random variable X we have $\mathbb{E}[|X|^3] \leq (\mathbb{E}X^4)^{3/4}$, so that

$$\mathbb{E}[|M_i - \mathbb{E}M_i|^3] = O((nr_n^d)^{9/2}),$$

uniformly over $i \leq k_n$. Therefore

$$k_n \max_{1 \leq i \leq k_n} \frac{\mathbb{E}[|M_i - \mathbb{E}M_i|^3]}{(n^{1/2}nr_n^d)^3} = O\left(r_n^{-d}(nr_n^d)^{3/2}n^{-3/2}\right) = O(r_n^{d/2})$$

so we have (3.6) as required. $\square$

Exercise 3.1. Prove Lemma 3.2.

Exercise 3.2. Suppose $n^2 r_n^d \to \infty$ as $n \to \infty$. By using Exercise 2.2, show that $(\mathcal{E}'_n - \mathbb{E}\mathcal{E}'_n)/\sqrt{\mathbb{E}\mathcal{E}'_n} \xrightarrow{\mathcal{D}} \mathcal{N}(0,1)$. Show also that $(\mathcal{E}_n - \mathbb{E}\mathcal{E}_n)/\sqrt{\mathbb{E}\mathcal{E}_n} \xrightarrow{\mathcal{D}} \mathcal{N}(0,1)$.

Exercise 3.3. Show that in the proof of Theorem 3.4, the second-order term in $\mathbb{E}[M_i^3]$ comes to $3\mathbb{E}M_i\mathbb{E}R_i$ as asserted in that proof.

4 The maximum degree

For $\lambda > 0$, let $\text{Po}(\lambda)$ denote a Poisson distributed random variable with parameter λ. In this section we shall use the following lemma (see Exercise 4.1.)

Lemma 4.1. *Suppose* $0 < \lambda < \mu$ *and* $X = \text{Po}(\lambda)$ *and* $Y = \text{Po}(\mu)$. *Prove that for any* $k \in \mathbb{N}$ *we have* $\mathbb{P}[X \geq k] \leq \mathbb{P}[Y \geq k]$.

Let Δ_n denote the maximum degree of vertices in $G(\mathcal{X}_n, r_n)$, and let Δ'_n denote the maximum degree in $G(\mathcal{P}_n, r_n)$. Given $k \in \mathbb{N}$ let $N_{\geq k}(n)$ (respectively, $N'_{\geq k}(n)$) denote the number of vertices of $G(\mathcal{X}_n, r_n)$ (respectively, $G(\mathcal{P}_n, r_n)$) of degree at least k. Also set $N_k(n) := N_{\geq k}(n) - N_{\geq k+1}(n)$ and $N'_k(n) := N'_{\geq k}(n) - N'_{\geq k+1}(n)$ (the number of vertices of degree exactly k).

First consider the sparse limit with $n r_n^d \to 0$. The next result says, among other things, that if this convergence to zero is not too slow, then the maximum degree remains bounded in probability. By "not too slow" we here mean decaying like a power of n.

Theorem 4.2. *Let* $k \in \mathbb{N}$. *Then:*

(i) If $n r_n^d = o(n^{-1/k})$, *then* $\mathbb{P}[\Delta'_n \geq k] \to 0$.

(ii) If $n r_n^d = \omega(n^{-1/k})$, *then* $\mathbb{P}[\Delta'_n \geq k] \to 1$.

(iii) If $n r_n^d \sim \alpha n^{-1/k}$ *for some* $\alpha \in (0, \infty)$, *then* $\lim_{n\to\infty} \mathbb{P}[\Delta'_n = k]$ *exists and lies in* $(0, 1)$.

It follows from this result that if r_n is such that if $n^{-1/k} \ll n r_n^d \ll n^{-1/(k+1)}$ then $\mathbb{P}[\Delta'_n = k] \to 1$. It also follows that if $n r_n^d = \Theta(n^{-1/k})$, then

$$\lim_{n\to\infty} \mathbb{P}[\Delta'_n \in \{k-1, k\}] = 1.$$

This is the so-called *two-point concentration* (or *focusing*) property of the distribution of Δ'_n in this limiting regime.

Recall that for random variables $(X_n)_{n \geq 1}$ and any constant c, we say X_n tends to c in probability, and write $X_n \xrightarrow{P} c$ as $n \to \infty$, if for all $\varepsilon > 0$ we have $\mathbb{P}[|X_n - c| > \varepsilon] \to 0$ (as $n \to \infty$).

We shall prove (i) using what is commonly known in the theory of random graphs as the *first moment method*, and prove (ii) using the *second moment method.* In short, the first moment method involves showing some sequence of random variables X_n satisfies $X_n \xrightarrow{P} 0$ by using Markov's inequality, while the second moment method involves showing some sequence Y_n satisfies $Y_n \xrightarrow{P} 1$ by using Chebyshev's inequality.

Proof of Theorem 4.2. Note that

$$\mathbb{P}[\text{Po}(\lambda) \geq k] \sim \lambda^k/k! \text{ as } \lambda \downarrow 0. \tag{4.1}$$

Therefore using the Mecke formula, if $nr_n^d \to 0$ we have

$$\mathbb{E}N'_{\geq k}(n) \sim n(n\theta r_n^d)^k/k! \text{ as } n \to \infty. \tag{4.2}$$

Suppose $nr_n^d = o(n^{-1/k})$. Then $n(nr_n^d)^k \to 0$. Hence by (4.2), $\mathbb{E}N'_{\geq k}(n) \to 0$, so by Markov's inequality $\mathbb{P}[N'_{\geq k}(n) \geq 1] \to 0$ and hence $\mathbb{P}[\Delta'_n \geq k] \to 0$. This gives us part (i).

Now suppose $nr_n^d \to 0$ but $nr_n^d = \omega(n^{-1/k})$. Then $n(nr_n^d)^k \to \infty$, and (4.2) gives us that $\mathbb{E}N'_{\geq k}(n) \to \infty$. Writing just $N'_{\geq k}$ for $N'_{\geq k}(n)$, we have that $(N'_{\geq k})_2$ is the number of ordered pairs (X, Y) of distinct points of $\mathcal{P}_n$ such that $\mathcal{P}_n(B_{r_n}(X) \setminus \{x\}) \geq k$ and $\mathcal{P}_n(B_{r_n}(Y) \setminus \{Y\}) \geq k$. Hence by the Mecke formula we have

$$\mathbb{E}[(N'_{\geq k})_2] = \int \int n^2 \mathbb{P}[\mathcal{P}_n^y(B(x; r_n)) \geq k, \mathcal{P}_n^x(B(y; r_n)) \geq k] dx dy$$

$$= I_1(n) + I_2(n),$$

where $\mathcal{P}_n^x$ denotes the point process $\mathcal{P}_n \cup \{x\}$, and where $I_1(n)$ denotes the integral restricted to those (x, y) such that $\|x - y\| > 2r_n$, while $I_2(n)$ denotes the integral restricted to those (x, y) such that $\|x - y\| \leq 2r_n$.

For distinct $x, y \in (0, 1)^d$, if $\|x - y\| > 2r_n$, then by Lemma 4.1 we have

$$\mathbb{P}[\mathcal{P}_n^y(B(x; r_n)) \geq k, \mathcal{P}_n^x(B(y; r_n)) \geq k] \leq (\mathbb{P}[\text{Po}(n\theta r_n^d) \geq k])^2,$$

with equality for large enough n. Hence by (4.1) and dominated convergence, we have as $n \to \infty$ that

$$I_1(n) \sim n^2((n\theta r_n^d)^k/k!)^2 \sim (\mathbb{E}N'_{\geq k})^2.$$

Also, $\mathbb{P}[\mathcal{P}_n^y(B(x; r_n)) \geq k, \mathcal{P}_n^x(B(y; r_n)) \geq k] \leq \mathbb{P}[\mathcal{P}_n(B(x; r_n)) \geq k - 1]$ for any x, y, so

$$I_2(n) \leq n^2\theta(2r_n)^d \mathbb{P}[\text{Po}(n\theta r_n^d) \geq k - 1] = O(n(nr_n^d)^k)$$

so that $I_2(n)/(\mathbb{E}N'_{\geq k})^2 = O(n^{-1}(nr_n^d)^{-k}) = o(1)$, and hence

$$\frac{\mathbb{E}[(N'_{\geq k})^2]}{(\mathbb{E}[N'_{\geq k}])^2} = \frac{\mathbb{E}[(N'_{\geq k})_2]}{(\mathbb{E}[N'_{\geq k}])^2} + \frac{\mathbb{E}N'_{\geq k}}{(\mathbb{E}[N'_{\geq k}])^2} \to 1.$$

Hence, $\text{Var}\left[N'_{\geq k}/\mathbb{E}N'_{\geq k}\right] \to 0$, and hence by Chebyshev's inequality $N'_{\geq k}/\mathbb{E}N'_{\geq k}$ $\xrightarrow{P} 1$, so in particular $\mathbb{P}[\Delta'_n \geq k] = \mathbb{P}[N'_{\geq k} > 0] \to 1$, completing the proof of (ii) under the extra assumption that $nr_n^d \to 0$.

In general, if $nr_n^d = \omega(n^{-1/k})$ then we can find a sequence $(s_n)_{n\in\mathbb{N}}$ with $n^{-1/k} \ll ns_n^d \ll 1$, and $s_n \leq r_n$ for all n; for example take $s_n = \min(r_n, n^{-(1/d)-1/(2kd)})$. Setting Δ''_n to be the maximum degree in $G(\mathcal{P}_n, s_n)$ we have that $\Delta''_n \leq \Delta'_n$, and $\mathbb{P}[\Delta''_n \geq k] \to 1$ by the above, completing the proof of (ii) in the general case.

We prove part (iii) only for $k = 1$ (see Reference [1, Corollary 6.5] for the general case). Suppose $nr_n^d \sim \alpha n^{-1}$, with $\alpha \in (0,1)$; then $n^2 r_n^d \to \alpha$, and $\mathbb{E}\mathcal{E}'_n \to \theta\alpha/2$ by (2.2). Then by Theorem 2.2 we have

$$\mathbb{P}[\Delta'_n = 0] = \mathbb{P}[\mathcal{E}'_n = 0] \to \exp(-\theta\alpha/2)$$

but also since $nr_n^d = o(n^{-1/2})$, by part (i) we have $\mathbb{P}[\Delta'_n \geq 2] \to 0$. Therefore $\lim_{n\to\infty} \mathbb{P}[\Delta'_n = 1] = 1 - e^{-\theta\alpha/2} \in (0,1)$. $\qquad\square$

Next, we briefly consider the thermodynamic limit with $nr_n^d \to \alpha$ for some $\alpha \in (0,\infty)$. The following result is immediate from Theorem 4.2 (ii).

Proposition 4.3. *Suppose* $\liminf nr_n^d > 0$. *Let* $k \in \mathbb{N}$. *Then* $\mathbb{P}[\Delta'_n \geq k] \to 1$.

Thus in the thermodynamic limit, and also (by Theorem 4.2 (ii)) in the case with $nr_n^d \to 0$ but $\liminf n^2 r_n^d > 0$, we have that $\Delta'_n \gg nr_n^d$ in probability.

Now we consider the dense limit; specifically the case where the degree of a typical vertex grows logarithmically in n. That is, we now assume

$$\frac{n\theta r_n^d}{\log n} \to \alpha, \quad \alpha \in (0,\infty). \tag{4.3}$$

In this case we shall find it is *not* the case that $\Delta'_n \gg nr_n^d$ in probability. We shall give a weak law showing that $\Delta'_n/(nr_n^d)$ tends to a positive constant in probability.

To state the result we shall need more notation. Define the function $H :$ $(1,\infty) \to \mathbb{R}$ by

$$H(a) = 1 - a + a\log a.$$

By elementary calculus, $H'(a) = \log a$ so $H(1) = 0$ is the unique minimum value of $H(\cdot)$, and $H(\cdot)$ is increasing on $(1,\infty)$ and decreasing on $(0,1)$; also $\lim_{a\downarrow 0} H(a) = 1$ and $\lim_{a\to\infty} H(a) = +\infty$.

For $x \geq 0$ let $H_+^{-1}(x)$ be the $a \in [1,\infty)$ with $H(a) = x$, and if $0 \leq x < 1$ let $H_-^{-1}(x)$ be the $a \in (0,1]$ with $H(a) = x$.

Theorem 4.4. *Suppose* (4.3) *holds. Then* $\Delta'_n/(n\theta r_n^d) \xrightarrow{P} H_+^{-1}(1/\alpha)$ *as* $n \to \infty$.

Remarks

1. It can be shown that for the random geometric graph in the d-dimensional unit *torus*, if (4.3) holds with $\alpha > 1$ and δ_n denotes the *minimum* degree of $G(\mathcal{P}_n, r_n)$ then $\delta_n/(n\theta r_n^d) \to H_-^{-1}(1/\alpha)$ in probability. See Exercise 4.3.
2. The statement of Theorem 4.4 also holds with Δ_n' replaced by Δ_n, and with convergence in probability $\xrightarrow{P}$ replaced by almost sure convergence. Moreover, if $nr_n^d/\log n \to \infty$ but $r_n \to 0$, then $\Delta_n/n\theta r_n^d \to 1$, almost surely. Proving these extensions is beyond the scope of these notes (see [1, Theorem 6.14]).

The function H arises from the following large-deviations results (Chernoff bounds) concerning the Poisson distribution.

Lemma 4.5. *Let* $\lambda > 0, t > 0$. *Then*

$$\mathbb{P}[\text{Po}(\lambda) \geq t] \leq \exp(-\lambda H(t/\lambda)), \quad t \geq \lambda; \tag{4.4}$$

$$\mathbb{P}[\text{Po}(\lambda) \leq t] \leq \exp(-\lambda H(t/\lambda)), \quad t \leq \lambda. \tag{4.5}$$

Also, for any $a > 1$ *and* $\varepsilon > 0$, *there exists* $\lambda_0 \in (0, \infty)$ *such that*

$$\mathbb{P}[\text{Po}(\lambda) \geq a\lambda] \geq \exp(-(1+\varepsilon)\lambda H(a)), \quad \lambda \geq \lambda_0. \tag{4.6}$$

Proof. Set $X = \text{Po}(\lambda)$. For $z \geq 1$, by Markov's inequality applied to the random variable z^X,

$$\mathbb{P}[X \geq t] = \mathbb{P}[z^X \geq z^t] \leq z^{-t}\mathbb{E}[z^X] = z^{-t}e^{\lambda(z-1)}. \tag{4.7}$$

For $z \leq 1$, similarly

$$\mathbb{P}[X \leq t] = \mathbb{P}[z^X \geq z^t] \leq z^{-t}\mathbb{E}[z^X] = z^{-t}e^{\lambda(z-1)}. \tag{4.8}$$

Put $z = t/\lambda$. If $t \geq \lambda$ then this choice of z satisfies $z \geq 1$ so by (4.7) we obtain that

$$\mathbb{P}[X \geq t] \leq \left(\frac{\lambda}{t}\right)^t e^{t-\lambda} = \exp(-\lambda H(t/\lambda)),$$

which proves (4.4). If $t \leq \lambda$ then the same choice of z satisfies $z \leq 1$ so we obtain (4.5) from (4.8).

Finally to prove (4.6), we use the following inequality (a weak form of Stirling's formula): for $k \in \mathbb{N}$ we have

$$\log k! = \sum_{i=1}^{k} \log i \leq \int_1^{k+1} \log x\, dx = (k+1)\log(k+1) - k$$

so $k! \leq (k+1)^{k+1}e^{-k}$. Thus if we fix $a > 1$ and put $k = \lceil a\lambda \rceil$ we obtain

$$\mathbb{P}[X \geq a\lambda] \geq \mathbb{P}[X = k] = e^{-\lambda}\lambda^k/k!$$

$$\geq \frac{e^{-\lambda}e^k\lambda^k}{(k+1)^{k+1}}$$

and hence

$$\lambda^{-1}\log\mathbb{P}[\mathrm{Po}(\lambda) \geq a\lambda] \geq -1 + (k/\lambda) - (k/\lambda)\log((k+1)/\lambda)$$

$$-\lambda^{-1}\log(k+1)$$

$$\geq -H(a)(1+\varepsilon), \quad \text{for } \lambda \text{ sufficiently large,}$$

which proves (4.6). $\qquad\qquad\qquad\qquad\qquad\qquad\qquad\qquad\qquad\qquad\Box$

Proof of Theorem 4.4. Assume that (4.3) holds, i.e. that $n\theta r_n^d/\log n \to \alpha$ with $\alpha \in (0,\infty)$. Let $\beta > H_+^{-1}(1/\alpha)$, so that $H(\beta) > 1/\alpha$. Let $\delta > 0$ be chosen so small that $(1+\delta)^d < \beta$ and

$$\alpha H(\beta/(1+\delta)^d) > 1+\delta. \qquad (4.9)$$

Cover $[0,1]^d$ by a deterministic collection of balls of radius δr_n, using as few balls as possible. The number of balls required, denoted k_n, is $O(r_n^{-d})$ and therefore is $O(n/\log n)$. Let the centres of these balls be denoted $x_{n,1}, \ldots, x_{n,k_n}$.

If $\Delta_n \geq n\theta r_n^d\beta$, then there is a point X of $\mathcal{P}_n$ with degree at least $n\theta r_n^d\beta$ in $G(\mathcal{P}_n, r_n)$, and this must lie in one of the balls $B(x_{n,i}; \delta r_n)$, $1 \leq i \leq k_n$, say for $i = I$. Then by the triangle inequality, we must have $\mathcal{P}_n(B(x_{n,I}; (1+\delta)r_n)) \geq n\theta r_n^d\beta$, and hence we have the event inclusion

$$\{\Delta_n \geq n\theta r_n^d\beta\} \subset \cup_{i=1}^{k_n}\{\mathcal{P}_n(B(x_{n,i}; (1+\delta)r_n)) \geq n\theta r_n^d\beta\}. \qquad (4.10)$$

Now for each $i \leq k_n$, the random variable $W_{n,i} := \mathcal{P}_n(B(x_{n,i}; (1+\delta)r_n))$ is Poisson distributed with mean satisfying

$$\mathbb{E}W_{n,i} \leq n\theta(1+\delta)^d r_n^d$$

so that by Lemma 4.1 and Lemma 4.5, for n large

$$\mathbb{P}[W_{n,i} \geq n\theta r_n^d\beta] \leq \exp\left[-n\theta(1+\delta)^d r_n^d H\left(\frac{n\theta r_n^d\beta}{n\theta(1+\delta)^d r_n^d}\right)\right]$$

$$\leq \exp\left[-\alpha(\log n)H\left(\frac{\beta}{(1+\delta)^d}\right)\right].$$

Then using (4.9) we have

$$\mathbb{P}[W_{n,i} \geq n\theta r_n^d\beta] \leq n^{-(1+\delta)}.$$

Therefore by (4.10) and the union bound,

$$\mathbb{P}[\Delta_n \geq n\theta r_n^d \beta] \leq \sum_{i=1}^{k_n} \mathbb{P}[W_{n,i} \geq n\theta r_n^d \beta] = O((n/\log n)n^{-(1+\delta)})$$

so that

$$\mathbb{P}[\Delta_n \geq n\theta r_n^d \beta] \to 0, \qquad \beta > H_+^{-1}(1/\alpha). \qquad (4.11)$$

Now suppose $1 < \gamma < H_+^{-1}(1/\alpha)$, so that $H(\gamma) < 1/\alpha$. Let $\delta > 0$, to be chosen below. Choose a maximal deterministic collection of points $y_{n,1}, \ldots, y_{n,j_n} \in [0,1]^d$ such that the balls $B(y_{n,i}; (1-\delta)r_n), 1 \leq i \leq j_n$ are disjoint and all contained in $[0,1]^d$. Then $j_n = \Theta(r_n^{-d}) = \Theta(n/\log n)$. Given n, for $1 \leq i \leq j_n$, define the event

$$A_{n,i} := \{\mathcal{P}_n(B(y_{n,i}; \delta r_n)) \geq 1\}$$
$$\cap \{\mathcal{P}_n(B(y_{n,i}; (1-\delta)r_n) \setminus B(y_{n,i}; \delta r_n)) \geq n\theta r_n^d \gamma\}.$$

Then we have the event inclusion

$$\cup_{i=1}^{j_n} A_{n,i} \subset \{\Delta_n' \geq n\theta r_n^d \gamma\}. \qquad (4.12)$$

Now, since $\delta < 1/2$ we have $\delta^d < 2\delta(1-\delta)^d$ (e.g. by induction on d). Hence $(1-\delta)^d - \delta^d > (1-\delta)^d(1-2\delta) > (1-2\delta)^{d+1}$, so $V_{n,i} := \mathcal{P}_n(B(y_{n,i}; (1-\delta)r_n) \setminus B(y_{n,i}; \delta r_n))$ is Poisson with mean

$$\mathbb{E}V_{n,i} = n\theta((1-\delta)^d - \delta^d)r_n^d > (1-2\delta)^{d+1}n\theta r_n^d;$$

also, for all large enough n we have $(1+\delta)\mathbb{E}V_{n,i} < \alpha \log n$ for $1 \leq i \leq j_n$. Hence by Lemma 4.5, for large enough n we have

$$\mathbb{P}[V_{n,i} \geq n\theta r_n^d \gamma] \geq \exp\left[-(1+\delta)(\mathbb{E}V_{n,i})H\left(\frac{n\theta r_n^d \gamma}{\mathbb{E}V_{n,i}}\right)\right]$$

$$\geq \exp\left[-\alpha(\log n)H\left(\frac{\gamma}{(1-2\delta)^{d+1}}\right)\right]$$

and if δ is chosen so small that $\alpha H\left(\frac{\gamma}{(1-2\delta)^{d+1}}\right) < 1 - \delta$, we obtain that

$$\mathbb{P}[V_{n,i} \geq n\theta r_n^d \gamma] \geq \exp(-(1-\delta)\log n) = n^{\delta-1}.$$

Since the event $\{\mathcal{P}_n(B(y_{n,i}; \delta r_n)) \geq 1\}$ has probability greater than $1/2$ (for large enough n) and is independent of the event $\{V_{n,i} \geq n\theta r_n^d \gamma\}$, we have $\mathbb{P}[A_{n,i}] \geq (1/2)n^{\delta-1}$. Therefore, since the events $A_{n,1}, \ldots, A_{n,j_n}$ are independent we have for some $c > 0$ that

$$\mathbb{P}[\cap_{i=1}^{j_n} A_{n,i}^c] \leq (1 - \frac{1}{2}n^{\delta-1})^{j_n} \leq \exp(-cn^{\delta-1} \times (n/\log n))$$

which tends to zero. Combined with (4.12), this shows that

$$\mathbb{P}[\Delta'_n < \gamma\, n\theta r_n^d] \to 0, \quad 1 < \gamma < H_+^{-1}(1/\alpha).$$

Combined with (4.11) this gives us the result. $\quad\square$

Exercise 4.1. Prove Lemma 4.1.

Exercise 4.2. Prove that for any $a \in (0,1)$ and $\varepsilon > 0$, there exists $\lambda_1 \in (0,\infty)$ such that, similarly to (4.6),

$$\mathbb{P}[\text{Po}(\lambda) \le a\lambda] \ge \exp(-(1+\varepsilon)\lambda H(a)), \quad \lambda \ge \lambda_1. \qquad (4.13)$$

Exercise 4.3. Prove the assertion in the first remark after Theorem 4.4.

5 A sufficient condition for connectivity

A graph (V, E) is said to be *connected* if there is a path in it from u to v for all distinct $u, v \in V$. A *component* of G is a maximal connected subgraph of G, i.e. a connected subgraph of G that is not a proper subgraph of any other connected subgraph of G.

Let $\mathcal{K}$ be the class of connected graphs. In this section we prove the following result.

Theorem 5.1. *Suppose $d = 2$ and $(r_n)_{n\in\mathbb{N}}$ is such that*

$$n\pi r_n^2 / \log n \to \alpha > 1. \qquad (5.1)$$

Then $\mathbb{P}[G(\mathcal{P}_n, r_n) \in \mathcal{K}] \to 1$.

Note that for the random geometric graph $G(\mathcal{P}_n, r_n)$ in the two-dimensional unit *torus*, the expected number of isolated vertices equals $n\exp(-n\pi r_n^2)$ by the Mecke formula (Lemma 2.3), and therefore tends to zero as $n \to \infty$ under the condition (5.1). Theorem 5.1 may be interpreted as saying that firstly, boundary effects are not important when considering isolated vertices in $d = 2$, and secondly, that the condition of having no isolated vertices (clearly necessary for $G(\mathcal{P}_n, r_n)$ to be connected) is also close to being sufficient. Moreover, our proof can (and will) be adapted to show that this condition is actually close to being sufficient for $G(\mathcal{P}_n, r_n)$ to satisfy stronger notions of connectivity, namely k-connectivity and Hamiltonicity as discussed below.

The proof of Theorem 5.1 is long, and requires a series of lemmas. It proceeds by discretization of space. Throughout this section we assume $d = 2$.

Assume r_n is given, satisfying (5.1). Let $\varepsilon \in (0, 1/9)$ be chosen in such a way that

$$\alpha(1 - \varepsilon)(1 - 9\varepsilon) > 1 + \varepsilon. \qquad (5.2)$$

Given $n \in \mathbb{N}$, set $\varepsilon_n := (r_n \lceil 1/(\varepsilon r_n) \rceil)^{-1}$. Then $\varepsilon_n \leq \varepsilon$ but $\varepsilon_n \to \varepsilon$ as $n \to \infty$, and $1/(\varepsilon_n r_n) \in \mathbb{N}$. Divide $[0,1]^2$ into squares of side $\varepsilon_n r_n$. Let $\mathcal{L}_n$ be the set of centers of these squares (a finite lattice). Then $|\mathcal{L}_n| = \Theta(n/\log n)$.

List the squares as $Q_{n,i}, 1 \leq i \leq |\mathcal{L}_n|$, and the corresponding centers of squares (i.e. the elements of $\mathcal{L}_n$) as $q_{n,i}, 1 \leq i \leq |\mathcal{L}_n|$.

Given $K \in \mathbb{N}$, let us say $q_{n,i} \in \mathcal{L}_n$ is K-*occupied* if $\mathcal{P}_n(Q_{n,i}) \geq K$. Let $\mathcal{O}_n^K$ be the (random) set of sites $q_{n,i} \in \mathcal{L}_n$ that are K-occupied.

Lemma 5.2. *If $G(\mathcal{P}_n, r_n)$ is disconnected, then so is $G(\mathcal{O}_n^1, r_n(1 - 2\varepsilon))$.*

Proof. If $q_{n,i}, q_{n,j} \in \mathcal{L}_n$ with $\|q_{n,i} - q_{n,j}\| \leq r_n(1 - 2\varepsilon)$, then for any $X \in \mathcal{P}_n \cap Q_{n,i}$ and $Y \in \mathcal{P}_n \cap Q_{n,j}$, by the triangle inequality we have

$$\|X - Y\| \leq \|X - q_{n,i}\| + \|q_{n,i} - q_{n,j}\| + \|Y - q_{n,j}\|$$

$$\leq r_n \varepsilon_n + r_n(1 - 2\varepsilon) + r_n \varepsilon_n = r_n$$

and therefore if $G(\mathcal{O}_n^1, r_n(1 - 2\varepsilon))$ is connected, so is $G(\mathcal{P}_n, r_n)$. □

Given nonempty sets $A, B \subset \mathbb{R}^2$ set $\mathrm{dist}(A, B) := \inf\{\|x - y\| : x \in A, y \in B\}$. Given also $z \in \mathbb{R}^2$, write just $\mathrm{dist}(z, B)$ for $\mathrm{dist}(\{z\}, B)$. Here ∂A denotes the boundary of A, i.e. the intersection of the closure of A and that of its complement.

Given $n, m \in \mathbb{N}$, let $\mathcal{A}_{n,m}$ denote the set of $\sigma \subset \mathcal{L}_n$ with m elements such that $G(\sigma, r_n(1 - 2\varepsilon))$ is connected (sometimes called "lattice animals").

Let $\mathcal{A}_{n,m}^2$ be the set of $\sigma \in \mathcal{A}_{n,m}$ such that $\mathrm{dist}(\sigma, \partial[0,1]^2) \geq 2r_n$, i.e. all elements of σ are distant at least $2r_n$ from the boundary of $[0,1]^2$. Let $\partial[0,1]^2$ denote the boundary of $[0,1]^2$.

Let $\mathcal{A}_{n,m}^1$ be the set of $\sigma \in \mathcal{A}_{n,m}$ such that σ is distant less than $2r_n$ from *just one edge* of $[0,1]^2$.

Let $\mathcal{A}_{n,m}^0 := \mathcal{A}_{n,m}^0 \setminus (\mathcal{A}_{n,m}^2 \cup \mathcal{A}_{n,m}^1)$, the set of $\sigma \in \mathcal{A}_{n,m}$ such that σ is distant less than $2r_n$ from *two edges* of $[0,1]^2$ (i.e. near a corner of $[0,1]^2$).

Lemma 5.3. *Given $m \in \mathbb{N}$, there is a constant $C = C(m)$ such that*

$$|\mathcal{A}_{n,m}| \leq C(n/\log n), \quad |\mathcal{A}_{n,m}^1| \leq C(n/\log n)^{1/2}, \quad |\mathcal{A}_{n,m}^0| \leq C$$

for all $n \in \mathbb{N}$, with $|\cdot|$ here denoting cardinality.

Proof. Fix m. Consider how many ways there are to choose $\sigma \in \mathcal{A}_{n,m}$.

There are at most r_n^{-2} choices, and hence $O(n/\log n)$ choices, for the first element of σ in the lexicographic ordering. Once the first element of σ is chosen, there are a bounded number of ways to choose the rest of σ.

Consider how many ways there are to choose $\sigma \in \mathcal{A}_{n,m}^1$. In this case there are $O(r_n^{-1}) = O((n/\log n)^{1/2})$ ways to choose the first element of σ (distant at most $2r_n$ from the boundary of $[0,1]^2$), and then a bounded number of ways to choose the rest of σ.

Finally consider how many ways there are to choose $\sigma \in \mathcal{A}_{n,m}^0$. In this case there are $O(1)$ ways to choose the first element of σ, and then a bounded number of ways to choose the rest of σ. $\quad\square$

Given any $K, n \in \mathbb{N}$ and any nonempty $\sigma \subset \mathcal{L}_n$ with $G(\sigma, r_n(1 - 2\varepsilon))$ connected, let E_σ^K be the event that σ is (the vertex-set of) a component of $G(\mathcal{O}_n^K, r_n(1 - 2\varepsilon))$. Let $\tilde{E}_\sigma^K$ be the event that there is no $v \in \mathcal{O}_n^K$ with $0 < \text{dist}(v, \sigma) \le r_n(1 - 2\varepsilon)$. Note that $E_\sigma^K = \tilde{E}_\sigma^K \cap \{\sigma \subset \mathcal{O}_n^K\}$.

In the following lemma, we use only the case $K = 1$ for the present section but will use other values of K for the next section.

Lemma 5.4. *Suppose that r_n satisfy (5.1) and $\varepsilon \in (0, 1/9)$ has been chosen to satisfy (5.2). Let $K, m \in \mathbb{N}$. Then there exists $n_0 \in \mathbb{N}$ such that for $n \ge n_0$ we have*

$$\sup_{\sigma \in \mathcal{A}_{n,m}^2} \mathbb{P}[\tilde{E}_\sigma^K] \le n^{-(1+\varepsilon)}, \tag{5.3}$$

and also

$$\sup_{\sigma \in \mathcal{A}_{n,m}^1} \mathbb{P}[\tilde{E}_\sigma^K] \le n^{-(1+\varepsilon)/2}, \tag{5.4}$$

and

$$\sup_{\sigma \in \mathcal{A}_{n,m}^0} \mathbb{P}[\tilde{E}_\sigma^K] \le n^{-(1+\varepsilon)/4}. \tag{5.5}$$

Proof. Given $\sigma \in \mathcal{A}_{n,m}^2$, let $q_{n,i}$ (respectively $q_{n,j}$) be the lexicographically first (resp. last) element of σ. Let B_σ^- be the part of $B(q_{n,i}; (1 - 3\varepsilon)r_n)$ lying strictly to the left of $Q_{n,i}$. Let B_σ^+ be the part of $B(q_{n,j}; (1 - 3\varepsilon)r_n)$ lying strictly to the right of $Q_{n,j}$.

Suppose $k \in \{1, \ldots, |\mathcal{L}_n|\}$ with $Q_{n,k} \cap B_\sigma^- \ne \emptyset$. Then $q_{n,k} \notin \sigma$, and by the triangle equality,

$$\|q_k - q_i\| \le (1 - 3\varepsilon)r_n + \varepsilon r_n = (1 - 2\varepsilon)r_n,$$

so if $\tilde{E}_\sigma^K$ occurs, then $\mathcal{P}_n(Q_{n,k}) < K$ for all such k. Likewise for B_σ^+. Hence,

$$\mathbb{P}[\tilde{E}_\sigma^K] \le \mathbb{P}[\cap_{\{k : Q_{n,k} \cap B_\sigma^- \ne \emptyset\} \cup \{k : Q_{n,k} \cap B_\sigma^+ \ne \emptyset\}} \{\mathcal{P}_n(Q_{n,k}) < K\}]. \tag{5.6}$$

For large enough n and for $1 \le k \le |\mathcal{L}_n|$ we have

$$\mathbb{P}[\mathcal{P}_n(Q_{n,k}) < K] = \sum_{j=0}^{K-1} e^{-\lambda_n} \lambda_n^j / j! \le K \lambda_n^K e^{-\lambda_n},$$

where we set $\lambda_n := n(\varepsilon_n r_n)^2$, so that by (5.1) we have $\lambda_n \sim \varepsilon^2 \alpha (\log n)/\pi$ as $n \to \infty$. Therefore, for large enough n we have

$$\mathbb{P}[\mathcal{P}_n(Q_{n,1}) < K] \leq n^{-\varepsilon^2(1-\varepsilon)\alpha/\pi}. \tag{5.7}$$

Now the total area of squares $Q_{n,k}$ such that $Q_{n,k} \cap B_\sigma^- \neq \emptyset$ or $Q_{n,k} \cap B_\sigma^+ \neq \emptyset$ is at least the area of $B_\sigma^- \cup B_\sigma^+$, which is at least $\pi((1-3\varepsilon)r_n)^2 - 2\varepsilon r_n^2$ and hence exceeds $\pi(1-8\varepsilon)r_n^2$. Therefore, for sufficiently large n, the number of such squares is at least $\pi(1-9\varepsilon)/\varepsilon^2$, and hence by (5.6) and (5.7) we obtain that

$$\mathbb{P}[\tilde{E}_\sigma^K] \leq n^{-\alpha(1-9\varepsilon)(1-\varepsilon)}.$$

By (5.2), this is less than $n^{-1-\varepsilon}$, completing the proof of (5.3).

To prove (5.4), take $\sigma \in \mathcal{A}_{n,m}^1$. Consider just the case where σ is near to the left edge of $[0,1]^2$. Define B_σ^+ as above. Similarly to (5.6) we have

$$\mathbb{P}[E_\sigma^K] \leq \mathbb{P}[\cap_{k:Q_{n,k} \cap B_\sigma^+ \neq 0} \{\mathcal{P}_n(Q_{n,k}) < K\}].$$

For n sufficiently large, the number of squares $Q_{n,k}$ such that $Q_{n,k} \cap B_\sigma^+ \neq \emptyset$ is at least $\pi(1-9\varepsilon)/(2\varepsilon^2)$, so by (5.7) and (5.2) we have

$$\mathbb{P}[E_\sigma^K] \leq n^{-\alpha(1-\varepsilon)(1-9\varepsilon)/2} \leq n^{-(1+\varepsilon)/2},$$

completing the proof of (5.4).

The proof of (5.5) is similar. See Exercise 5.1. $\quad\square$

The *order* of a graph is the number of vertices it has. For $m \in \mathbb{N}$, let $\mathcal{K}_m$ be the class of graphs having at least one component of order m.

Lemma 5.5. *Let* $K, m \in \mathbb{N}$. *Then* $\mathbb{P}[G(\mathcal{O}_n^K, (1-2\varepsilon)r_n) \in \mathcal{K}_m] \to 0$ *as* $n \to \infty$.

Proof. Since $E_\sigma^K \subset \tilde{E}_\sigma^K$ for all σ, by Lemma 5.4 we have for large enough n that

$$\mathbb{P}[G(\mathcal{O}_n^K, (1-2\varepsilon)r_n) \in \mathcal{K}_m] \leq \sum_{\sigma \in \mathcal{A}_{n,m}} \mathbb{P}(\tilde{E}_\sigma^K)$$

$$\leq |\mathcal{A}_{n,m}^2| n^{-(1+\varepsilon)} + |\mathcal{A}_{n,m}^1| n^{-(1+\varepsilon)/2} + |\mathcal{A}_{n,m}^0| n^{-(1+\varepsilon)/4}$$

and using Lemma 5.3 we find that this tends to zero. $\quad\square$

Lemma 5.6. (Unicoherence of $[0,1]^2$; see [1, Lemma 9.1]) *For any two closed connected subsets* A, B *of* $[0,1]^2$ *with union* $A \cup B = [0,1]^2$, *the intersection* $A \cap B$ *is connected.*

Proof. Omitted. $\quad\square$

Given $K, L \in \mathbb{N}$, let $F_L^K(n)$ be the event that $G(\mathcal{O}_n^K, (1 - 2\varepsilon)r_n)$ has two or more components of order L or more. In other words, set

$$F_L^K := \cup_{\ell, m \geq L} \cup_{\sigma \in A_{n,\ell}} \cup_{\{\tau \in A_{n,m} : \tau \cap \sigma = \emptyset\}} (E_\sigma^K \cap E_\tau^K). \tag{5.8}$$

For $m \in \mathbb{N}$, let $A'_{n,m}$ be the set of $*$-connected subsets of $\mathcal{L}_n$ with m elements, where we say a set $\tau \subset \mathcal{L}_n$ is $*$-connected if for any two distinct sites x, y in τ, there is a path $(x_0, x_1, \ldots, x_k)$ with $x_0 = x$, $x_k = y$ and $\|x_i - x_{i-1}\|_\infty = \varepsilon_n r_n$ for $1 \leq i \leq k$.

The next lemma says that for reasonably large L, if F_L^K occurs then there exists a fairly large $*$-connected set in $\mathcal{L}_n$ for which the corresponding squares contain few points of $\mathcal{P}_n$.

Lemma 5.7. *Suppose* $K, L \in \mathbb{N}$ *with* $L \geq 64$. *Then there exists* $n_0 \in \mathbb{N}$ *such that for* $n \geq n_0$ *we have*

$$F_L^K(n) \subset \cup_{m \geq \sqrt{L}/4} \cup_{\tau \in A'_{n,m}} \cap_{i : q_{n,i} \in \tau} \{\mathcal{P}_n(Q_{n,i}) < K\}. \tag{5.9}$$

Proof. Choose n_0 so that $\varepsilon_n \geq \varepsilon/2$ for $n \geq n_0$. From now on assume $n \geq n_0$. Suppose $F_L^K(n)$ occurs. Then there exist $U \subset \mathcal{O}_n^K$, $V \subset \mathcal{O}_n^K$, such that $\min(|U|, |V|) \geq L$ and $G(U, (1 - 2\varepsilon)r_n)$ and $G(V, (1 - 2\varepsilon)r_n)$ are distinct components of $G(\mathcal{O}_n^K, (1 - 2\varepsilon)r_n)$. Let

$$U' := [0, 1]^2 \cap \left(\cup_{q \in U} B(q; (1/2)(1 - 2\varepsilon)r_n) \right);$$
$$V' := [0, 1]^2 \cap \left(\cup_{q \in V} B(q; (1/2)(1 - 2\varepsilon)r_n) \right).$$

Then U', V' are connected, closed, disjoint subsets of $[0, 1]^2$. Let $\tilde{V}$ be the closure of the component of $[0, 1]^2 \setminus U'$, containing V', and let $\tilde{U}$ be the closure of $[0, 1]^2 \setminus \tilde{V}$; essentially, this is the set obtained by filling in the holes of U' not connected to V'.

Then $\tilde{U}, \tilde{V}$ are closed connected sets, whose union is $[0, 1]^2$, and whose intersection (the "exterior boundary" of U' with respect to V') is therefore connected by Lemma 5.6.

Let $\partial U'$ denote the intersection of U' with the closure of $[0, 1]^2 \setminus U'$. We assert that

$$\tilde{U} \cap \tilde{V} \subset \partial U'. \tag{5.10}$$

To see this, suppose $x \in [0, 1]^2 \setminus U'$, then (a) if x lies in the component of $[0, 1]^2 \setminus U'$ containing V', then x lies in the interior of $\tilde{V}$ (relative to $[0, 1]^2$), and hence $x \notin \tilde{U}$, and (b) if x lies in some other component of $[0, 1]^2 \setminus U'$, then x lies in the interior of this component (relative to $[0, 1]^2$), and hence $x \notin \tilde{V}$. Now suppose x lies in the interior of U' (relative to $[0, 1]^2$). Then x is not in the closure of $[0, 1]^2 \setminus U'$, and so $x \notin \tilde{V}$. Thus we have (5.10).

Let τ be the set of $q_{n,i}$ such that $Q_{n,i} \cap (\tilde{U} \cap \tilde{V}) \neq \emptyset$. Then τ is $*$-connected in $\mathcal{L}_n$.

For each $q_{n,i} \in \tau$ we claim $\mathcal{P}_n(Q_{n,i}) < K$. Indeed, by (5.10) any such $Q_{n,i}$ contains part of $\partial U'$, so if $\mathcal{P}_n(Q_{n,i}) \geq K$ then $q_{n,i} \in \mathcal{O}_n^K$ and $q_{n,i}$ is distant at most $(1/2)r_n$ from U and therefore $q_{n,i}$ would actually be in U so $Q_{n,i}$ would *not* include any of $\partial U'$, which is a contradiction.

We claim that τ satisfies the isoperimetric inequality

$$|\tau| \geq (1/4)L^{1/2}, \tag{5.11}$$

where $|\cdot|$ here denotes cardinality, from which we may deduce (5.9).

To see (5.11), for bounded nonempty $A \subset \mathbb{R}^2$ let $\mathrm{diam}_\infty(A) := \max\{\|x - y\|_\infty : x, y \in A\}$, where $\|(x_1, x_2)\|_\infty := \max(|x_1|, |x_2|)$ for $(x_1, x_2) \in \mathbb{R}^2$. Then since τ is $*$-connected,

$$\mathrm{diam}_\infty(\tilde{U} \cap \tilde{V}) \leq \mathrm{diam}_\infty(\tau) + \varepsilon_n r_n \leq \varepsilon r_n(|\tau| + 1).$$

Let S be a rectilinear square of side $\varepsilon r_n(|\tau| + 1)$ containing the set $\tilde{U} \cap \tilde{V}$.

Then S also contains either U or V; if not we can find $u \in U \setminus S$ and $v \in V \setminus S$, and a continuous path in $[0,1]^2 \setminus S$ from U to V, but then this path would pass through a point of $\tilde{U} \cap \tilde{V}$, which is a contradiction.

Assume without loss of generality that $U \subset S$. Since the union of squares of side $\varepsilon_n r_n$ centered on points of U is contained in the square of side $\varepsilon r_n(|\tau| + 2)$ with the same center as S, since $n \geq n_0$ we have

$$(\varepsilon r_n(|\tau| + 2))^2 \geq |U|(\varepsilon_n r_n)^2 \geq L(\varepsilon r_n/2)^2,$$

and hence $L \leq 4(|\tau| + 2)^2$. Since our assumption $L \geq 64$ implies $\sqrt{L}/2 - 2 \geq \sqrt{L}/4$, this gives us (5.11). $\square$

Lemma 5.8. *Let $K \in \mathbb{N}$. There exists $L \in \mathbb{N}$ such that $\mathbb{P}[F_L^K(n)] \to 0$ as $n \to \infty$.*

Proof. We claim that there are finite constants γ and C such that for $n, m \in \mathbb{N}$ we have

$$|\mathcal{A}'_{n,m}| \leq C(n/\log n)\gamma^m. \tag{5.12}$$

To see this, observe that the number of $*$-connected subsets of $\mathbb{Z}^2$ with m elements containing the origin is bounded by 256^m (see [1, Lemma 9.3]). As in the proof of Lemma 5.3, there are $O(n/\log n)$ ways to choose the first element (in the lexicographic ordering) of a subset of $\mathcal{L}_n$, and then at most 256^m ways to choose the rest of the subset, demonstrating (5.12) with $\gamma = 256$. Set

$$\phi_n := \mathbb{P}[\mathcal{P}_n(Q_{n,1}) < K] \leq K(n(\varepsilon_n r_n)^2)^K \exp(-n(\varepsilon_n r_n)^2)$$

$$= \exp[-\varepsilon^2(\alpha/\pi)(\log n)(1 + o(1))],$$

where the last line comes from (5.1). Then for $L \geq 64$, by (5.9) and (5.12) we have

$$\mathbb{P}[F_L^K(n)] \leq \sum_{m \geq \sqrt{L}/4} C(n/\log n)\gamma^m \phi_n^m$$

$$\leq 2Cn(\gamma\, n^{-\varepsilon^2/\pi})^{\sqrt{L}/4}$$

which tends to zero provided L is chosen so that $\varepsilon^2 L^{1/2} > 4\pi$ (as well as $L \geq 64$). $\quad\square$

Lemma 5.9. *Let $K \in \mathbb{N}$. Then as $n \to \infty$ we have*

$$\mathbb{P}[G(\mathcal{O}_n^K, (1-2\varepsilon)r_n) \notin \mathcal{K}] \to 0.$$

Proof. Choose $L \in \mathbb{N}$ as in Lemma 5.8. Then

$$\mathbb{P}[G(\mathcal{O}_n^K, (1-2\varepsilon)r_n) \notin \mathcal{K}]$$

$$\leq \left(\sum_{m=1}^{L} \mathbb{P}[G(\mathcal{O}_n^K, (1-2\varepsilon)r_n) \in \mathcal{K}_m] \right) + \mathbb{P}[F_L^K(n)].$$

By Lemmas 5.5 and 5.8, this tends to zero. $\quad\square$

Proof of Theorem 5.1. By Lemmas 5.2 and 5.9 we have

$$\mathbb{P}[G(\mathcal{P}_n, r_n) \notin \mathcal{K}] \leq \mathbb{P}[G(\mathcal{O}_n^1, r_n(1-2\varepsilon)) \notin \mathcal{K}] \to 0. \quad\square$$

Exercise 5.1. Prove (5.5).

Exercise 5.2. Assume $d = 2$ and $(r_n)_{n \in \mathbb{N}}$ satisfies (5.1). Let $K, L \in \mathbb{N}$ with $L \geq 64$. Define the event

$$\tilde{F}_L^K := \cup_{\ell, m \geq L} \cup_{\sigma \in \mathcal{A}_{n,\ell}} \cup_{\{\tau \in \mathcal{A}_{n,m}: \mathrm{dist}(\tau,\sigma) > r_n(1-2\varepsilon)\}} (\tilde{E}_\sigma^K \cap \tilde{E}_\tau^K).$$

Show, similarly to Lemma 5.7, that there exists $n_0 \in \mathbb{N}$ such that for $n \geq n_0$ we have

$$\tilde{F}_L^K(n) \subset \cup_{m \geq \sqrt{L}/4} \cup_{\tau \in \mathcal{A}'_{n,m}} \cap_{i: q_{n,i} \in \tau} \{\mathcal{P}_n(Q_{n,i}) < K\}. \tag{5.13}$$

Exercise 5.3. Let $k \in \mathbb{N}$. A graph G with more than $k + 1$ vertices is said to be k-vertex-connected if for any two vertices there are at least k vertex-disjoint paths between them. Equivalently, it is said to be k-vertex-connected if there is no way to remove k vertices that disconnects the graph. This equivalence is a consequence of *Menger's theorem*; see Penrose [1] for a statement of Menger's theorem.

Let $\tilde{\mathcal{K}}_k$ be the class of k-vertex-connected graphs. Prove that under the hypothesis of Theorem 5.1, we have $\mathbb{P}[G(\mathcal{P}_n, r_n) \in \tilde{\mathcal{K}}_k] \to 1$.

6 Connectivity and Hamiltonicity

A *Hamiltonian cycle* in a finite graph $G = (V, E)$ is an enumeration $x_1, x_2, \ldots, x_n$ of V such that, setting $x_0 = x_n$, one has $\{x_{i-1}, x_i\} \in E$ for $i = 1, \ldots, n$. A graph possessing at least one Hamiltonian cycle is said itself to be *Hamiltonian*. Let $\mathcal{H}$ be the set of all Hamiltonian graphs, and define the *Hamiltonicity threshold*

$$H_n' = \min\{r : G(\mathcal{P}_n, r) \in \mathcal{H}\},$$

which is a random variable determined by the configuration of $\mathcal{P}_n$. Similarly, let $\mathcal{K}$ be the class of connected graphs, and define the *connectivity threshold* to be the random variable

$$\rho_n' = \min\{r : G(\mathcal{P}_n, r) \in \mathcal{K}\}.$$

We may also define

$$H_n = \min\{r : G(\mathcal{X}_n, r) \in \mathcal{H}\}; \qquad \rho_n = \min\{r : G(\mathcal{X}_n, r) \in \mathcal{K}\}.$$

In this section we prove the following result.

Theorem 6.1. *Assume $d = 2$ (so $\theta = \pi$). Then*

$$n\theta(\rho_n')^2 / \log n \xrightarrow{P} 1, \tag{6.1}$$

and also

$$n\theta(H_n')^2 / \log n \xrightarrow{P} 1. \tag{6.2}$$

Remarks

(i) The restriction to $d = 2$ arises because boundary effects become more important in higher dimensions (and $d = 1$ is different because 1-space is "less connected").

(ii) The convergence (6.1) actually holds with almost sure convergence, and these hold with ρ_n' replaced by ρ_n, but proving these extensions is beyond the scope of these notes. See Penrose [1].

(iii) A further extension of (6.1) is the following convergence in distribution result: for $d = 2$, and for any $t \in \mathbb{R}$,

$$\lim_{n \to \infty} \mathbb{P}[n\theta(\rho_n')^2 - \log n \leq t] = \exp(-e^{-t}).$$

Proving this is also beyond our scope here. Again, see Reference [1].

Given r_n, let δ_n' denote the minimum degree of $G(\mathcal{P}_n, r_n)$. Given any event A, let A^c denote its complement.

Theorem 6.2. *If $n\theta r_n^d / \log n \to \alpha < 1$ then $\mathbb{P}[\delta_n' = 0] \to 1$.*

Proof. Let $\varepsilon > 0$ be chosen so that $\alpha(1 + \varepsilon)^d < 1 - 2\varepsilon$. Given n, choose a maximal collection of points $x_{n,i} \in [0,1]^2, 1 \le i \le k_n$ such that the balls $B_{n,i}^+ := B(x_{n,i}; (1+\varepsilon)r_n), 1 \le i \le k_n$ are disjoint and contained in $[0,1]^d$. Note that $k_n = \Theta(n / \log n)$.

Then let $B_{n,i}^- := B(x_{n,i}; \varepsilon r_n)$, for $1 \le i \le k_n$, and define the events

$$E_{n,i} := \{ \mathcal{P}_n(B_{n,i}^+) = \mathcal{P}_n(B_{n,i}^-) = 1 \}, \quad 1 \le i \le k_n.$$

If $\cup_{i=1}^{k_n} E_{n,i}$ occurs then $\delta_n' = 0$. Writing $|\cdot|$ for Lebesgue measure, we have for large n that

$$\begin{aligned}
\mathbb{P}[E_{n,i}] &= e^{-n|B_{n,i}^-|}(n|B_{n,i}^-|) \times e^{-n|B_{n,i}^+ \setminus B_{n,i}^-|} \\
&= \exp(-n\theta((1+\varepsilon)r_n)^d) \times n\theta(\varepsilon r_n)^d \\
&\ge \exp(-(1-\varepsilon)\log n) \times \Theta(\log n) = \Theta(n^{\varepsilon-1}\log n),
\end{aligned}$$

and so there is a constant $\delta > 0$ such that

$$\begin{aligned}
\mathbb{P}[\cap_{i=1}^{k_n} E_{n,i}^c] &\le (1 - \delta n^{\varepsilon-1}\log n)^{k_n} \le \exp(-k_n \delta n^{\varepsilon-1}\log n) \\
&\le \exp(-\Theta(n^\varepsilon)) \to 0.
\end{aligned}$$

Therefore $\mathbb{P}[\delta_n' = 0] \ge \mathbb{P}[\cup_{i=1}^{k_n} E_{n,i}] \to 1$, as claimed. $\square$

Corollary 6.3. *Given $\varepsilon > 0$, we have $\mathbb{P}[n\theta(\rho_n')^d / \log n > 1 - \varepsilon] \to 1$, and $\mathbb{P}[n\theta(H_n')^d / \log n > 1 - \varepsilon] \to 1$.*

Proof. Assume $\varepsilon < 1$. Set $r_n = ((1 - \varepsilon)\log n / (n\theta))^{1/d}$, so $n\theta r_n^d / \log n = 1 - \varepsilon$. Let δ_n' be the minimum degree of $G(\mathcal{P}_n, r_n)$. If the minimum degree of a graph of order greater than 1 is zero, then it is not connected; hence

$$\begin{aligned}
\mathbb{P}[n\theta(\rho_n')^d / \log n \le 1 - \varepsilon] &= \mathbb{P}[G(\mathcal{P}_n, r_n) \in \mathcal{K}] \\
&\le \mathbb{P}[\delta_n' > 0] + \mathbb{P}[\mathcal{P}_n([0,1]^d) \le 1],
\end{aligned}$$

which tends to zero by Theorem 6.2.

If a graph is Hamiltonian, then clearly it is connected. Therefore $H_n' \ge \rho_n'$, so we also have as $n \to \infty$ that

$$\mathbb{P}[n\theta(H_n')^d / \log n > 1 - \varepsilon] \to 1. \quad \square$$

The first part (6.1) of Theorem 6.1, is immediate from Theorem 5.1 and Corollary 6.3.

Given a finite graph $G = (V, E)$ and a function $f : V \to \mathbb{N}$, define the *f-blowup* of G to be the graph with vertex set $\cup_{v \in V}\{(v, i) : i \in \{1, 2, \dots, f(v)\}\}$, with an edge from (v, i) to (w, j) if and only if either $v = w$ or $\{v, w\}$ is an edge of G. The proof of (6.2) in Theorem 6.1 uses the following lemma.

Lemma 6.4. *Suppose $G = (V, E)$ is a finite connected graph with at least two vertices and $f : V \to \mathbb{N}$ is a function with $f(v) \geq \text{Deg}(v)$ for all $v \in V$. Then the f-blowup of G is Hamiltonian.*

Proof. It suffices to prove the result in the case where G is a *tree*. We proceed by induction on $|V|$; clearly the result holds for all trees (V, E) with $|V| = 2$.

Let $n \in \mathbb{N}$ with $n \geq 2$, and suppose the result holds for all trees (V, E) with $|V| = n$. Let $G = (V, E)$ be a tree with $|V| = n + 1$, and suppose $f : V \to \mathbb{N}$ with $f(v) \geq \text{Deg}(v)$ for all $v \in V$.

Pick a leaf v of G (i.e. a vertex of degree 1), and take $w \in G$ with $\{v, w\} \in E$. Let $G' = (V', E')$ be the graph G with vertex v and edge $\{v, w\}$ removed, and let $f' : V' \to \mathbb{N}$ be given by $f'(w) = f(w) - 1$ and $f'(u) = f(u)$ for all $u \in V' \setminus \{w\}$.

By the inductive hypothesis, the f'-blowup of G' is Hamiltonian. Let $x_1, \ldots, x_m$ be a Hamiltonian cycle through this graph, chosen so that $x_m = (w, 1)$.

Put $x_{m+j} = (v, j)$ for $1 \leq j \leq f(v)$, and $x_{m+f(v)+1} = (w, f(w))$. Then $(x_1, \ldots, x_{m+f(v)+1})$ is a Hamiltonian cycle through the f-blowup of G so that the f-blowup of G is Hamiltonian. This completes the induction. $\square$

To complete the proof of (6.2) in Theorem 6.1, it suffices to prove the following.

Theorem 6.5. *Suppose $d = 2$ and $(r_n)_{n \in \mathbb{N}}$ is such that*

$$n\theta r_n^2 / \log n \to \alpha > 1. \tag{6.3}$$

Then $\mathbb{P}[G(\mathcal{P}_n, r_n) \in \mathcal{H}] \to 1$.

Proof. Assume $d = 2$ and r_n is given, satisfying (6.3). Let $\varepsilon \in (0, 1/9)$ be chosen in such a way that

$$(1 - \varepsilon)\alpha(1 - 9\varepsilon) > 1 + \varepsilon. \tag{6.4}$$

As in the preceding section, divide $[0, 1]^2$ into squares of side $\varepsilon_n r_n$ with $\varepsilon_n = (r_n \lceil 1/(\varepsilon r_n) \rceil)^{-1}$. Let $\mathcal{L}_n$ be the set of centers of these squares. List the squares as $Q_{n,i}, 1 \leq i \leq |\mathcal{L}_n|$, and the corresponding centers of squares (i.e. the elements of $\mathcal{L}_n$) as $q_{n,i}, 1 \leq i \leq |\mathcal{L}_n|$. Let $\mathcal{A}_{n,m}, \mathcal{A}_{n,m}^2, \mathcal{A}_{n,m}^1$ and $\mathcal{A}_{n,m}^0$ be as defined in the preceding section.

Given $K \in \mathbb{N}$, let $\mathcal{O}_n^K$ be the (random) set of sites $q_{n,i} \in \mathcal{L}_n$ such that $\mathcal{P}_n(Q_{n,i}) \geq K$.

Now take $K = \lceil (99/\varepsilon)^2 \rceil$. By Lemma 5.9, $\mathbb{P}[A_n] \to 1$ as $n \to \infty$, where we set

$$A_n := \{G(\mathcal{O}_n^K, (1 - 2\varepsilon)r_n) \in \mathcal{K}\}.$$

Also, by the case $m = 1$ of Lemmas 5.3 and 5.4, we have $\mathbb{P}[B_n] \to 1$, where we set

$$B_n := \cap_{1 \le i \le |\mathcal{L}_n|} \{ \mathrm{dist}(q_{n,i}, \mathcal{O}_n^K) \le (1 - \varepsilon) r_n \}. \tag{6.5}$$

We claim that if n is such that $\varepsilon_n \ge \varepsilon/2$ (which holds for all but finitely many n), and $A_n \cap B_n$ occurs, then $G(\mathcal{P}_n, r_n)$ is Hamiltonian. To justify this, assume now that $A_n \cap B_n$ occurs; it suffices to prove that the f-blowup of $G(\mathcal{O}_n^1, (1 - 2\varepsilon)r_n)$ is Hamiltonian, where we put $f(q_{n,i}) = \mathcal{P}_n(Q_{n,i})$. This suffices because $G(\mathcal{P}_n, r_n)$ has a subgraph (with the same vertex set) that is isomorphic to the f-blowup of $G(\mathcal{O}_n^1, (1 - 2\varepsilon)r_n)$, namely the subgraph of $G(\mathcal{P}_n, r_n)$ obtained by retaining all edges connecting vertices X, Y such that $X \in Q_{n,i}$ and $Y \in Q_{n,j}$ for some i, j with $\|q_{n,i} - q_{n,j}\| \le r_n(1 - 2\varepsilon)$, and discarding all other edges.

Take a spanning subgraph G' of $G(\mathcal{O}_n^1, (1 - 2\varepsilon)r_n)$ as follows. For each vertex $v \in \mathcal{O}_n^1 \setminus \mathcal{O}_n^K$, choose a vertex $w \in \mathcal{O}_n^K$ with $0 < \|w - v\| \le (1 - 2\varepsilon)r_n$ (which is possible because we assume B_n occurs), retain the edge from v to w but discard all edges from v except the one to w. Retain all edges which are between two vertices in $\mathcal{O}_n^K$.

Since we assume A_n occurs, $G(\mathcal{O}_n^K, (1 - 2\varepsilon)r_n)$ is connected, and hence so is G'. Moreover, $f(v) \ge \mathrm{Deg}(v)$ for each vertex v of G'; this holds by construction for $v \in \mathcal{O}_n^1 \setminus \mathcal{O}_n^K$, and holds for other v because the vertex degrees in $G(\mathcal{L}_n, (1 - 2\varepsilon)r_n)$ are all bounded by $(2r_n/(\varepsilon_n r_n))^2$, and hence by K, assuming $\varepsilon_n \ge \varepsilon/2$.

Therefore by Lemma 6.4 the f-blowup of G' is Hamiltonian, and therefore so is the f-blowup of $G(\mathcal{O}_n^1, (1 - 2\varepsilon)r_n)$, justifying our claim. The result follows. $\square$

Exercise 6.1. For a finite point set $\mathcal{X} \subset \mathbb{R}^d$ with at least two elements, the *largest nearest neighbor link* of $\mathcal{X}$ is defined as

$$\mathrm{LNNL}(\mathcal{X}) := \max_{x \in \mathcal{X}} \min_{y \in \mathcal{X} \setminus \{x\}} (\|y - x\|);$$

as a matter of convention, set $\mathrm{LNNL}(\mathcal{X}) = +\infty$ for $|\mathcal{X}| \le 1$. Show that the statement of Theorem 6.1 also holds with ρ_n' replaced by $\mathrm{LNNL}(\mathcal{P}_n)$.

Exercise 6.2. Let $k \in \mathbb{N}$. Let $\rho_n^{(k)}$ be the minimum r such that $G(\mathcal{P}_n, r)$ is k-vertex-connected. Show that for $d = 2$, (6.1) holds with ρ_n' replaced by $\rho_n^{(k)}$.

7 Solutions to exercises

Ex. 1.1 Suppose $A_1, \ldots, A_k$ form a partition of $[0, 1]^d$. Given $N_n = m$, the distribution of $(\mathcal{P}_n(A_1), \ldots, \mathcal{P}_n(A_k))$ is multinomial with parameters $(m; |A_1|, \ldots, |A_k|)$. Therefore for any $j_1, \ldots, j_k \in \mathbb{N} \cup \{0\}$, setting $m =$

$\sum_{i=1}^{k} j_i$ we have

$$\mathbb{P}[\mathcal{P}_n(A_1) = j_1, \ldots, \mathcal{P}_n(A_k) = j_k]$$

$$= \left(\frac{e^{-n}n^m}{m!} \right) \times \frac{m!}{j_1! \cdots j_k!} \prod_{i=1}^{k} |A_i|^{j_i}$$

$$= \prod_{i=1}^{k} \frac{e^{-n|A_i|}(n|A_i|)^{j_i}}{j_i!}$$

so the variables $\mathcal{P}_n(A_i), 1 \leq i \leq k$, are independent Poisson distributed with parameter $n|A_i|$, demonstrating the result.

Ex. 1.2 Set $D = \text{Degree}(\xi_1)$. Conditional on ξ_1, the distribution of D is binomial with parameters $n - 1$ and $|B(\xi_1; r_n) \cap [0,1]^d|$, where for $x \in \mathbb{R}^d$ and $r > 0$, we let $B(x; r)$ denote the Euclidean ball of radius r centered at x. So by the law of total probability,

$$(nr_n^d)^{-1}\mathbb{E}D = \int_{[0,1]^d} n^{-1} r_n^{-d} \mathbb{E}[D|\xi_1 = x] dx$$

$$= \left(\frac{n-1}{n} \right) \int_{[0,1]^d} \left(\frac{|B(x; r_n) \cap [0,1]^d|}{r_n^d} \right) dx,$$

and the integrand tends to θ for almost all $x \in [0,1]^d$ and is bounded above by θ. So by dominated convergence this tends to θ, giving the result.

Ex. 2.1 We have that $\mathcal{E}_n = \sum_{1 \leq i < j \leq n} X_{ij}$ where X_{ij} is the indicator of the event that $\|\xi_i - \xi_j\| \leq r_n$, and therefore

$$\mathbb{E}\mathcal{E}_n = \binom{n}{2} \mathbb{P}[\|\xi_1 - \xi_2\| \leq r_n]$$

$$= \binom{n}{2} \int_{[0,1]^d} |B(x; r_n) \cap B(0; 1)| dx \qquad (7.1)$$

$$\sim \frac{n^2 r_n^d}{2} \int_{[0,1]^d} \left(\frac{|B(x; r_n) \cap B(0; 1)|}{r_n^d} \right) dx$$

and as in the previous question the integral tends to θ by dominated convergence, so (2.1) follows.

For (2.2) we use the Mecke formula (Lemma 2.3) with $k = 2$ and $f(x, y) = \mathbf{1}\{|x - y| \leq r_n\}$, to obtain

$$\mathbb{E}\mathcal{E}'_n = (1/2)\mathbb{E} \sum_{X, Y \in \mathcal{P}_n}^{\neq} f(X, Y)$$

$$= (1/2)n^2 \int_{[0,1]^d} |B(x; r_n) \cap B(0; 1)| dx$$

and then (2.2) follows by the same argument as was used to obtain (2.1) from (7.1).

Ex. 2.2 Following the hint, divide $[0,1]^d$ into cubes of side $1/m$, denoted $Q_1 \ldots, Q_{m^d}$, with the center of Q_i denoted q_i. Fix n and set

$$V_m = \{\{i,j\} : 1 \le i < j \le m^d \text{ and } \|q_i - q_j\| \le r_n\}.$$

For $\alpha = \{i,j\} \in V_m$ let W_α be the indicator of the event that $\mathcal{P}_n(Q_i) \times \mathcal{P}_n(Q_j) > 0$. Set

$$Z_m = \sum_{\alpha \in V_m} W_\alpha.$$

Then almost surely

$$\mathcal{E}'_n = \lim_{m \to \infty} Z_m.$$

Also $Z_m \le (\mathcal{P}_n([0,1]^d))^2$, so by dominated convergence we have as $m \to \infty$ (with n fixed) that $\mathbb{E}Z_m \to \mathbb{E}\mathcal{E}'_n = \lambda'_n$.

For $\alpha \ne \beta$ set $\alpha \sim \beta$ if and only if $\alpha \cap \beta \ne \emptyset$. Then by the independence properties of the Poisson process (V, α) is a dependency graph for $(W_\alpha, \alpha \in V_m)$.

Given n and m, for $\alpha, \beta \in V_m$ set $p_\alpha = \mathbb{E}W_\alpha$ and $p_{\alpha\beta} = \mathbb{E}W_\alpha W_\beta$. Then p_α does not depend on α and for fixed n we have $p_\alpha \sim (nm^{-d})^2$ as $m \to \infty$. Also $p_{\alpha\beta}$ does not depend on (α, β) (provided $\alpha \sim \beta$), and for fixed n we have $p_{\alpha\beta} \sim (nm^{-d})^3$ as $m \to \infty$. Thus, as $m \to \infty$ we have

$$\sum_{\alpha \in V_m} p_\alpha^2 = O(m^{2d} \times (n^2 m^{-2d})^2) = o(1),$$

and

$$\limsup_{m \to \infty} \sum_{\alpha,\beta \in V_m, \beta \sim \alpha} p_\alpha p_\beta \le \limsup_{m \to \infty} m^{3d} (\theta r_n^d)^2 \times (n^2 m^{-2d})^2 = 0,$$

and

$$\limsup_{m \to \infty} \sum_{\alpha,\beta \in V_m, \beta \sim \alpha} p_{\alpha\beta} \le \limsup_{m \to \infty} m^{3d} (\theta r_n^d)^2 \times (n^3 m^{-3d})$$

$$= \theta^2 n^3 r_n^{2d} \le cn r_n^d \lambda'_n,$$

for some constant c, where we set $\lambda'_n := \mathbb{E}\mathcal{E}'_n$, which is given asymptotically by (2.2). Therefore by Lemma 2.1,

$$\sum_{k=0}^{\infty} |\mathbb{P}[\mathcal{E}'_n = k] - e^{-\lambda'_n} (\lambda'_n)^k / k!| \le \min(2/\lambda'_n, 6) \times (c\lambda'_n n r_n^d + o(1))$$

which is $O(n r_n^d)$.

Ex. 3.1 We have

$$
\mathbb{E}[(X)_k] = \sum_{j=k}^{\infty} (j)_k \times e^{-\lambda} \lambda^j/j!
$$

$$
= \sum_{j=k}^{\infty} e^{-\lambda} \lambda^{k+(j-k)}/(j-k)!
$$

$$
= \lambda^k \sum_{i=0}^{\infty} e^{-\lambda} \lambda^i/i! = \lambda^k.
$$

Ex. 3.2 Suppose $nr_n^d \to 0$ and $n^2 r_n^d \to \infty$. Let $x \in \mathbb{R}^d$. Let Y_n be Poisson with parameter $\mathbb{E}\mathcal{E}_n'$ (which tends to ∞ as $n \to \infty$). Then by the central limit theorem $(Y_n - \mathbb{E}\mathcal{E}_n')/\sqrt{\mathbb{E}\mathcal{E}_n'}$ is asymptotically standard normal, i.e.

$$
\mathbb{P}\left[(Y_n - \mathbb{E}\mathcal{E}_n')/\sqrt{\mathbb{E}\mathcal{E}_n'} \le x\right] \to \Phi(x), \quad x \in \mathbb{R}.
$$

Also, for $x \in \mathbb{R}$ we have

$$
\left| \mathbb{P}\left[\frac{\mathcal{E}_n' - \mathbb{E}\mathcal{E}_n'}{\sqrt{\mathbb{E}\mathcal{E}_n'}} \le x \right] - \mathbb{P}\left[\frac{Y_n - \mathbb{E}\mathcal{E}_n'}{\sqrt{\mathbb{E}\mathcal{E}_n'}} \le x \right] \right|
$$

$$
\le \sup_{t \in \mathbb{R}} \left| \mathbb{P}[\mathcal{E}_n' \le t] - \mathbb{P}[Y_n \le t] \right|
$$

which tends to zero by Exercise 2.2 since for all t we have

$$
\mathbb{P}[\mathcal{E}_n' \le t] - \mathbb{P}[Y_n \le t] \le \sum_{m=0}^{\infty} |\mathbb{P}[\mathcal{E}_n' = m] - \mathbb{P}[Y_n = m]|.
$$

Hence $(\mathcal{E}_n' - \mathbb{E}\mathcal{E}_n')/\sqrt{\mathbb{E}\mathcal{E}_n'} \xrightarrow{\mathcal{D}} \mathcal{N}(0,1)$. The same result for $\mathcal{E}_n$ is proved similarly, using Theorem 2.2 instead of Exercise 2.2.

Ex. 3.3 Using notation from the proof of Theorem 3.4, we have that the second-order term in $\mathbb{E}[M_i^3]$ comes from triples of edges e, e', e'', all with left endpoint in C_i and with five distinct endpoints among them. There are three ways to decide which two of the three edges share a common endpoint, so the second-order term in $\mathbb{E}[M_i^3]$ comes to

$$
3\mathbb{E} \sum_{X,Y,Z,X_1,Y_1 \in \mathcal{P}_n}^{\neq} (g_{i,n}(X,Y) + g_{i,n}(Y,X))
$$

$$
\times (g_{i,n}(X,Z) + g_{i,n}(Z,X)) g_{i,n}(X_1,Y_1)
$$

and by the Mecke formula this comes to

$$3n^5 \int \cdots \int (g_{i,n}(x,y) + g_{i,n}(y,x))(g_{i,n}(x,z) + g_{i,n}(z,x))$$
$$\times g_{i,n}(x_1,y_1)dx\,dy\,dz\,dx_1\,dy_1$$

which equals $3\mathbb{E}R_i\mathbb{E}M_i$ by (3.10) and (3.8), as required.

Ex. 4.1 Note that Y has the same distribution as $X + Z$, where Z is Poisson with parameter $\mu - \lambda$, independent of X. Hence for any $k \in \mathbb{N}$ we have $\mathbb{P}[Y \geq k] = \mathbb{P}[X + Z \geq k] \geq \mathbb{P}[X \geq k]$.

Ex. 4.2 Putting $k = \lfloor a\lambda \rfloor$, and using the weak form of Stirling's formula as in the proof of Lemma 4.5, we obtain that

$$\mathbb{P}[\mathrm{Po}(\lambda) \leq a\lambda] \geq \mathbb{P}[\mathrm{Po}(\lambda) = k] = e^{-\lambda}\lambda^k/k!$$
$$\geq \frac{e^{-\lambda}e^k\lambda^k}{(k+1)^{k+1}}$$

and hence

$$\lambda^{-1}\log\mathbb{P}[\mathrm{Po}(\lambda) \leq a\lambda] \geq -1 + (k/\lambda) - (k/\lambda)\log((k+1)/\lambda)$$
$$-\lambda^{-1}\log(k+1)$$
$$\geq -H(a)(1+\varepsilon), \quad \text{for } \lambda \text{ sufficiently large,}$$

which proves the result.

Ex. 4.3 Suppose $n\theta r_n^d/\log n \to \alpha$ with $\alpha > 1$. We need to show that $\delta_n/(n\theta r_n^d) \to H_-^{-1}(1/\alpha)$ in probability, where for this question δ_n denotes the minimum degree of $G(\mathcal{P}_n, r_n)$ in the *torus*.

Choose β with $1 > \beta > H_-^{-1}(1/\alpha)$. Then $H(\beta) < 1/\alpha$. Let $\delta \in (0,1)$ be a constant (to be chosen below).

For each $n \in \mathbb{N}$, take a maximal collection of disjoint balls contained in $[0,1]^d$ of the form $B_{n,i}^+ := B(x_{n,i}; r_n(1+\delta))$, $1 \leq i \leq k_n$, with $k_n = \Theta(r_n^{-d}) = \Theta(n/\log n)$. Set $B_{n,i}^- := B(x_{n,i}; r_n\delta)$. Define the event

$$A_{n,i} := \{\mathcal{P}_n(B_{n,i}^-) = 1\} \cap \{\mathcal{P}_n(B_{n,i}^+ \setminus B_{n,i}^-) \leq n\theta r_n^d\beta\}.$$

Then $\mathbb{P}[A_{n,i}] = \mathbb{P}[A_{n,1}]$ for $1 \leq i \leq k_n$, and by Exercise 4.2

$$\mathbb{P}[A_{n,1}] \geq e^{-n|B_{n,i}^-|}(n|B_{n,i}^-|)$$
$$\times \exp\left(-(1+\delta)H\left(\frac{n\theta r_n^d\beta}{n\theta r_n^d((1+\delta)^d - \delta^d)}\right)n\theta r_n^d[(1+\delta)^d - \delta^d]\right).$$

Using our assumptions that $n\theta r_n^d / \log n \to \alpha$ and $\delta < 1$, we therefore have for large enough n that

$$\mathbb{P}[A_{n,1}] \geq e^{-2\delta\theta\alpha \log n}$$
$$\times \exp\left(-(1+\delta)^{d+2}H\left(\frac{\beta}{(1+\delta)^d - \delta^d}\right)\alpha \log n\right).$$

Suppose δ was chosen so that

$$(1+\delta)^{d+2}\alpha H\left(\frac{\beta}{(1+\delta)^d - \delta^d}\right) + 2\delta\theta\alpha < 1 - \delta.$$

Then $\mathbb{P}[A_{n,1}] \geq \exp(-(1-\delta)\log n) = n^{\delta-1}$. Therefore

$$\mathbb{P}[\cup_{i=1}^{k_n} A_{n,i}] \geq 1 - (1 - n^{\delta-1})^{k_n} \geq 1 - \exp(-k_n n^{\delta-1})$$

which tends to 1. But $\cup_{i=1}^{k_n} A_{n,i} \subset \{\delta_n \leq n\theta r_n^d \beta\}$, and therefore

$$\mathbb{P}[\delta_n \leq n\theta r_n^d \beta] \to 1, \quad 1 > \beta > H_-^{-1}(1/\alpha). \tag{7.2}$$

Now suppose that $0 < \gamma < H_-^{-1}(1/\alpha)$, so that $H(\gamma) > 1/\alpha$. Let $\eta > 0$ (to be chosen below). Cover the torus $[0,1]^2$ by balls of the form $C_{n,i}^- := B(y_{n,i}; \eta r_n), 1 \leq i \leq m_n$, with $m_n = \Theta(n/\log n)$.

For $n, i \in \mathbb{N}$ with $i \leq m_n$, define event $A_{n,i}' := \{\mathcal{P}_n(C_{n,i}^+) \leq n\theta r_n^d \gamma + 1\}$, where we put $C_{n,i}^+ := B(y_{n,i}; (1-\eta)r_n)$. Then $\{\delta_n \leq n\theta r_n^d \gamma\} \subset \cup_{i=1}^{m_n} A_{n,i}'$. Since we are on the torus, $\mathbb{P}[A_{n,i}'] = \mathbb{P}[A_{n,1}']$ for $1 \leq i \leq m_n$, and by Lemma 4.5, for n large enough

$$\mathbb{P}[A_{n,1}'] \leq \exp\left(-n\theta r_n^d(1-\eta)^d H\left(\frac{n\theta r_n^d \gamma + 1}{n\theta r_n^d(1-\eta)^d}\right)\right)$$
$$\leq \exp\left(-\alpha \log n(1-\eta)^{d+1}H\left(\frac{\gamma}{(1-\eta)^d} + \eta\right)\right)$$

and if η was chosen so that $\alpha(1-\eta)^{d+1}H\left(\frac{\gamma}{(1-\eta)^d} + \eta\right) > 1 + \eta$, then we have $\mathbb{P}[A_{n,1}'] \leq n^{-1-\eta}$, and hence

$$\mathbb{P}[\delta_n \leq n\theta r_n^d \gamma] \leq \mathbb{P}[\cup_{i=1}^{m_n} A_{n,i}'] \leq n \times n^{-1-\eta}$$

which tends to zero. Combined with (7.2), this completes the proof.

Ex. 5.1 To prove (5.5), take $\sigma \in \mathcal{A}_{n,m}^0$. Consider just the case where σ is near to the lower left corner of $[0,1]^2$. Define B_σ^+ as in the proof of Theorem 5.4 and let B_σ^{++} denote the upper half of B_σ^+. Similarly to (5.6) we have

$$\mathbb{P}[\tilde{E}_\sigma^K] \leq \mathbb{P}[\cap_{k:Q_{n,k} \cap B_\sigma^{++} \neq \emptyset} \{\mathcal{P}_n(Q_{n,k}) < K\}].$$

For n sufficiently large, the number of squares $Q_{n,k}$ such that $Q_{n,k} \cap B_\sigma^+ \neq \emptyset$ is at least $\pi(1 - 9\varepsilon)/(4\varepsilon^2)$, so by (5.7) and (5.2) we have

$$\mathbb{P}[\tilde{E}_\sigma^K] \leq n^{-\alpha(1-\varepsilon)(1-9\varepsilon)/4} \leq n^{-(1+\varepsilon)/4},$$

completing the proof of (5.5).

Ex. 5.2 Suppose $\tilde{F}_L^K$ occurs. Then there exist $U, V \in \cup_{m \geq L} \mathcal{A}_{n,m}$ with $\text{dist}(U, V) > r_n(1 - 2\varepsilon)$ and $\tilde{E}_U^K$ and $\tilde{E}_V^K$ both occurring. Assume n is large enough so that $\varepsilon_n \geq \varepsilon/2$. Define the set τ as in the proof of Lemma 5.7. Then by (5.10), for each $q_{n,i} \in \tau$, we have that $q_{n,i}$ is distant at most $(1/2)r_n$ from U but is also distant at least $((1/2) - 2\varepsilon)r_n$ from U so not in U itself. Therefore $\mathcal{P}_n(Q_{n,i}) < K$ for all such i, because we assume $\tilde{E}_U^K$ occurs.

Since τ is $*$-connected and (5.11) holds by the same arguments as in the proof of Lemma 5.7, the result (5.13) follows.

Ex. 5.3 Let $d = 2$ and assume $(r_n)_{n \in \mathbb{N}}$ satisfy (5.1). Suppose $G(\mathcal{P}_n, r_n)$ has more than k vertices but is not k-vertex-connected; then there is a set $\mathcal{C} \subset \mathcal{P}_n$ with at most k elements, such that $G(\mathcal{P}_n \setminus \mathcal{C}, r_n)$ is disconnected. Choose such a set $\mathcal{C}$ in an arbitrary way, and let $\tilde{\mathcal{O}}_n$ be the set of sites $q_{n,i} \in \mathcal{L}_n$ such that $(\mathcal{P}_n \setminus \mathcal{C})(Q_{n,i}) \geq 1$. Then by the proof of Lemma 5.2 the graph $G(\tilde{\mathcal{O}}_n, r_n(1 - 2\varepsilon))$ is disconnected.

Given nonempty $\sigma \subset \mathcal{L}_n$, we claim that if σ is a component of $G(\tilde{\mathcal{O}}_n, r_n(1 - 2\varepsilon))$, then the event $\tilde{E}_\sigma^{k+1}$ (defined just before Lemma 5.4) must occur. This is because for every $q_{n,i} \in \mathcal{L}_n \setminus \sigma$ with $\text{dist}(q_{n,i}, \sigma) \leq r_n(1 - 2\varepsilon)$, if $\mathcal{P}_n(Q_{n,i}) \geq k + 1$ then we must have $(\mathcal{P}_n \setminus \mathcal{C})(Q_{n,i}) \geq 1$, and therefore $q_{n,i} \in \tilde{\mathcal{O}}_n$, contradicting the assumption that σ is a component of $G(\tilde{\mathcal{O}}_n, r_n(1 - \varepsilon))$.

Therefore, defining $\tilde{F}_L^K(n)$ as in Exercise 5.2, for any $L \in \mathbb{N}$ we must have that either event $\tilde{F}_L^{k+1}(n)$ or (for some $m \leq L$ and $\sigma \in \mathcal{A}_{n,m}$) event $\tilde{E}_\sigma^{k+1}$ occurs.

Now, by Lemmas 5.3 and 5.4, for any fixed m we have

$$\mathbb{P}[\cup_{\sigma \in \mathcal{A}_{n,m}} \tilde{E}_\sigma^{k+1}] \leq \sum_{\sigma \in \mathcal{A}_{n,m}} \mathbb{P}[\tilde{E}_\sigma^{k+1}] \to 0,$$

and by the proof of Lemma 5.8, using (5.13) instead of (5.9), we can choose L so that $\mathbb{P}[\tilde{F}_L^{k+1}(n)] \to 0$. Combining these estimates gives us the result.

Ex. 6.1 Given any sequence r_n, we have $\text{LNNL}(\mathcal{P}_n) > r_n$ if and only if $\delta_n' = 0$, where δ_n' denotes the minimum degree of $G(\mathcal{P}_n, r_n)$ (or zero if $\mathcal{P}_n = \emptyset$) and therefore by Theorem 6.2, given $\alpha < 1$ we have

$$\mathbb{P}\left[\text{LNNL}(\mathcal{P}_n) > \left(\frac{\alpha \log n}{n\theta}\right)^{1/d}\right] \to 1. \tag{7.3}$$

On the other hand, a graph with more than one vertex must be free of isolated vertices if it is connected, and therefore $\text{LNNL}(\mathcal{P}_n) \leq \rho_n'$, except when $|\mathcal{P}_n| \leq 1$. Hence when $d = 2$, for $\beta > 1$ we have

$$\mathbb{P}\left[\mathrm{LNNL}(\mathcal{P}_n) > \left(\frac{\beta \log n}{n\theta}\right)^{1/2}\right]$$

$$\leq \mathbb{P}\left[\rho_n' < \left(\frac{\beta \log n}{n\theta}\right)^{1/2}\right] + \mathbb{P}[|\mathcal{P}_n| \leq 1]$$

which tends to zero by Theorem 6.1. Together with (7.3), this gives us the result.

Ex. 6.2 Clearly $\rho_n^{(k)} \geq \rho_n'$, so given $\varepsilon \in (0,1)$ we have by Corollary 6.3 that $\mathbb{P}[n\theta(\rho_n^{(k)})^d / \log n > 1 - \varepsilon] \to 1$. Also by Exercise 5.3, we have $\mathbb{P}[n\theta(\rho_n^{(k)})^d / \log n < 1 + \varepsilon] \to 1$. Together these give us the result.

References

[1] Penrose, M., *Random Geometric Graphs*, Oxford University Press, Oxford, 2003.
[2] Chen, L. and Shao, Q.-M., Normal approximation under local dependence, *Ann. Probab.* **32** (2004) 1985–2028.

4

On random graphs from a minor-closed class

Colin McDiarmid

1 Introduction

Let $\mathcal{G}$ be a class of (simple) graphs closed under isomorphism, for example the class $\mathcal{P}$ of planar graphs. Let $\mathcal{G}_n$ denote the set of graphs in $\mathcal{G}$ on vertex set $[n] := \{1, \ldots, n\}$. We write $R_n \in_u \mathcal{G}$ to mean that R_n is sampled uniformly at random from $\mathcal{G}_n$ (with the implicit assumption that this set is not empty).

What are typical properties of R_n? Is there usually a giant component? What is the asymptotic probability of being connected? Are there many leaves? How big is the 2-core? Denise et al. [1]. We are mainly interested in typical properties but to investigate these we need to consider counting.

Generating functions
Given a class $\mathcal{G}$ of graphs, the *exponential generating function* or *egf* is

$$G(x) = \sum_{n \geq 0} |\mathcal{G}_n| x^n / n! = \sum_{G \in \mathcal{G}} x^{v(G)} / v(G)!$$

where we divide by $n!$ since we are dealing with labeled graphs, see for example Flajolet and Sedgewick [2]. (Here $v(G)$ denotes the number of vertices in G.) Let $\rho_{\mathcal{G}}$ denote the radius of convergence of $G(x)$. Thus $0 \leq \rho_{\mathcal{G}} \leq \infty$, and

$$\rho_{\mathcal{G}}^{-1} = \limsup_{n \to \infty} (|\mathcal{G}_n| / n!)^{1/n}.$$

We will be most interested in classes $\mathcal{G}$ with $0 < \rho_{\mathcal{G}} < \infty$.

For a suitable graph class, we can interrelate the egfs (or two variable versions of them) for all graphs, connected graphs, 2-connected graphs and 3-connected graphs. If we know enough about the 3-connected graphs (as we do for example for planar graphs, thanks to the work of Tutte and others on maps) then we *may* be able to extend to all graphs.

Gimenéz and Noy [3] in 2009 extended earlier work of Bender et al. [4] to handle planar graphs. Chapuy et al. [5] and Bender and Gao [6] went on to handle the class $\mathcal{G}^S$ of graphs embeddable in any given surface S. Boltzmann sampling can also yield such results, but we do not pursue this here, see Chapter 2. We shall proceed in greater generality.

2 Properties of graph classes

2.1 Minor-closed classes

A graph H is a *minor* of G if H can be obtained from a subgraph of G by edge-contractions. The class $\mathcal{G}$ is *minor-closed* if

$$G \in \mathcal{G}, H \text{ a minor of } G \quad \Rightarrow \quad H \in \mathcal{G}$$

Examples of minor-closed classes include:

- forests, series-parallel graphs and more generally graphs of treewidth at most k;
- outerplanar graphs, planar graphs and more generally graphs embeddable on a given surface;
- graphs with at most k vertex-disjoint cycles.

We denote the class of graphs with no minor (isomorphic to) H by $\mathrm{Ex}(H)$. Similarly, given a set $\mathcal{H}$ of graphs, $\mathrm{Ex}(\mathcal{H})$ denotes the class of graphs with no minor graph in $\mathcal{H}$. Clearly any such class is minor-closed. For example, the class of series-parallel graphs is $\mathrm{Ex}(K_4)$, the class $\mathcal{P}$ of planar graphs is $\mathrm{Ex}(\{K_5, K_{3,3}\})$ and the class of graphs with no two vertex-disjoint cycles is $\mathrm{Ex}(2C_3)$. Here $2C_3$ denotes the disjoint union of two copies of the 3-cycle C_3.

It is easy to see that $\mathcal{G}$ is minor-closed if and only if $\mathcal{G} = \mathrm{Ex}(\mathcal{H})$ for some class $\mathcal{H}$. Indeed, if $\mathcal{G}$ is minor-closed then we may take $\mathcal{H}$ to consist of the graphs H which are not in $\mathcal{G}$ but each minor other than H itself is in $\mathcal{G}$: this is the unique minimal such $\mathcal{H}$, and we call the graphs in this set the *excluded minors* for $\mathcal{G}$. By Robertson and Seymour's fundamental graph minors theorem (once Wagner's conjecture), if $\mathcal{G}$ is minor-closed then it is $\mathrm{Ex}(\mathcal{H})$ for some **finite** class $\mathcal{H}$ (Robertson and Seymour [7]), or see for example Diestel [8]. For example, we noted that $\mathcal{P} = \mathrm{Ex}(\{K_5, K_{3,3}\})$, so K_5 and $K_{3,3}$ are the excluded minors for $\mathcal{P}$.

Mostly we shall assume that $\mathcal{G}$ is minor-closed and *proper* (that is, not empty and not all graphs). For such a class $\mathcal{G}$, there is a constant $c = c(\mathcal{G})$ such that the average degree of each graph in $\mathcal{G}$ is at most c, see Mader [9]. Thus our graphs are *sparse*. For $\mathrm{Ex}(K_t)$ the maximum average degree is of order $t\sqrt{\log t}$ (Kostochka [10], Thomason [11]).

Directly from the definition, the radius of convergence $\rho_{\mathcal{G}}$ is > 0 if and only if there is a constant c such that $|\mathcal{G}_n| \leq c^n n!$ for each n. Norine et al. [12] and

Dvořák and Norine [13] showed that each proper minor-closed graph class $\mathcal{G}$ has $\rho_{\mathcal{G}} > 0$.

2.2 Decomposable, bridge-addable and addable graph classes

When a graph is in $\mathcal{G}$ if and only if each component is, then we call the class $\mathcal{G}$ *decomposable*. For example, the class of planar graphs is decomposable but the class of graphs embeddable on the torus is not (consider the graph $2K_5$, consisting of two disjoint copies of K_5).

Proposition 2.1. *A minor-closed class $\mathcal{G}$ is decomposable if and only if each excluded minor is connected,*

Proof. If some excluded minor H is disconnected, then each component of H is in $\mathcal{G}$ but H is not in $\mathcal{G}$, so $\mathcal{G}$ is not decomposable. Conversely, let $\mathcal{H}$ be the set of excluded minors of $\mathcal{G}$, and suppose each graph in $\mathcal{H}$ is connected. If G is not in $\mathcal{G}$ then G has a minor $H \in \mathcal{H}$, so some component G' of G must have a minor H and thus $G' \notin \mathcal{G}$. It follows that $\mathcal{G}$ is decomposable. □

A *bridge* in a graph is an edge e such that deleting e increases the number of components. The class $\mathcal{G}$ is *bridge-addable* if whenever $G \in \mathcal{G}$ and u and v are vertices in different components of G then $G + uv \in \mathcal{G}$. Here $G + uv$ denotes the graph obtained from G by adding the edge uv. We say $\mathcal{G}$ is *addable* if it is both decomposable and bridge-addable. Recall that $\mathcal{G}^S$ denotes the class of graphs embeddable in a given surface S. The surface class $\mathcal{G}^S$ is bridge-addable but **not** decomposable except in the planar case.

Proposition 2.2. *A proper minor-closed class $\mathcal{G}$ is addable if and only if each excluded minor is 2-connected.*

Proof. Suppose $\mathcal{G}$ is addable. We want to show that each excluded minor H is 2-connected. Since $\mathcal{G}$ is decomposable, H must be connected. Suppose that H can be obtained from the disjoint union $H' \cup H''$ (neither consisting of a single vertex) by identifying a vertex v' in H' with a vertex v'' in H''. By minimality, both H' and H'' are in $\mathcal{G}$, and so $H' \cup H''$ is in $\mathcal{G}$. But then the graph obtained by adding the edge $v'v''$ to $H' \cup H''$ is in $\mathcal{G}$, and contracting this edge shows that H is in $\mathcal{G}$, a contradiction.

Conversely, suppose $\mathcal{G}$ is not addable. We want to show that some excluded minor is not 2-connected. If $\mathcal{G}$ is not decomposable then it has a disconnected excluded minor, so we may assume that $\mathcal{G}$ is not bridge-addable. Thus there are graphs G' and G'' in $\mathcal{G}$ such that the graph G obtained by adding an edge between a vertex of G' and a vertex of G'' is not in $\mathcal{G}$. Now G must contain an excluded minor H. But since G' and G'' are in $\mathcal{G}$, H must have a cut vertex. □

3 Bridge-addability, being connected and the fragment

It is perhaps natural to expect that a typical graph in a bridge-addable class should not have many components, as there are many ways to join components together. Similarly, we might expect almost all vertices to be in one "giant" component. Let us first consider the probability of there being just one component.

3.1 Bridge-addability and being connected

The following non-asymptotic lower bound on the probability of being connected is from McDiarmid et al. in 2005 [14].

Theorem 3.1. *If the class $\mathcal{G}$ is bridge-addable and $R_n \in_u \mathcal{G}$ then*

$$\mathbb{P}(R_n \text{ is connected}) \geq 1/e.$$

To prove this result we use one preliminary lemma. Given a graph G, let comp(G) denote the number of components, let Bridge(G) denote the set of bridges and let Cross(G) denote the set of "cross-edges" or "possible edges" between components.

Lemma 3.2. *If the graph G has n vertices, then $|\text{Bridge}(G)| \leq n - \text{comp}(G)$; and if* comp$(G) = k+1$ *then*

$$|\text{Cross}(G)| \geq k(n-k) + \binom{k}{2} \geq k(n-k).$$

Proof. If H is a component of G and T is a spanning tree of H, then Bridge$(H) \subseteq E(T)$ and $e(T) = v(H) - 1$. Summing over components gives the first inequality. Now consider the second inequality, and assume that comp$(G) = k+1$ for a positive integer k. Since if $0 < |X| \leq |Y|$ then $|X||Y| > (|X| - 1)(|Y| + 1)$, we may see that $|\text{Cross}(G)|$ is minimized when G consists of k singleton components and one other component. $\qquad \square$

Proof of Theorem 3.1. Let $\mathcal{G}_n^k$ denote the set of graphs in $\mathcal{G}_n$ with k components. The key observation is that, if $G \in \mathcal{G}_n^{k+1}$ and $e \in \text{Cross}(G)$, then $G' = G + e$ is in $\mathcal{G}_n^k$ and $e \in \text{Bridge}(G')$. Hence, using also Lemma 3.2, for $1 \leq k \leq n-1$,

$$|\mathcal{G}_n^k| \cdot (n-k) \geq |\{(G',e) : G' \in \mathcal{G}_n^k, e \in \text{Bridge}(G')\}|$$
$$\geq |\{(G,e) : G \in \mathcal{G}_n^{k+1}, e \in \text{Cross}(G)\}|$$
$$\geq |\mathcal{G}_n^{k+1}| \cdot k(n-k).$$

Thus

$$|\mathcal{G}_n^{k+1}| \leq \frac{1}{k}|\mathcal{G}_n^k|. \tag{3.1}$$

Write C_n for the class $\mathcal{G}_n^1$ of connected graphs in $\mathcal{G}_n$. By using the last bound repeatedly we see that $|\mathcal{G}_n^{k+1}| \leq \frac{1}{k!}|C_n|$ for each $k = 1, 2, \ldots$. Thus

$$|\mathcal{G}_n| = \sum_{k=0}^{n-1} |\mathcal{G}_n^{k+1}| \leq |C_n| \sum_{k=0}^{n-1} \frac{1}{k!} < |C_n| \cdot e,$$

yielding $|C_n|/|\mathcal{G}_n| > 1/e$.   □

For trees $\mathcal{T}$ and forests $\mathcal{F}$, the classical result of Cayley says that $|\mathcal{T}_n| = n^{n-2}$, and Rényi showed in 1959 [15] that $|\mathcal{F}_n| \sim e^{\frac{1}{2}} n^{n-2}$, see also [16–18]. Thus for $F_n \in_u \mathcal{F}$,

$$\mathbb{P}(F_n \text{ is connected}) = \frac{|\mathcal{T}_n|}{|\mathcal{F}_n|} \sim e^{-\frac{1}{2}} \approx 0.6065.$$

(For comparison, observe that $e^{-1} \approx 0.3679$.) This result was noted in McDiarmid et al. [19], and it was conjectured there that if $\mathcal{G}$ is bridge-addable then

$$\mathbb{P}(R_n \text{ is connected}) \geq e^{-\frac{1}{2}+o(1)}.$$

Balister et al. [20, 21] gave an asymptotic lower bound of $e^{-0.7983}$. Norine improved this to $e^{-2/3}$ in an unpublished work in 2013. Under the extra condition that $\mathcal{G}$ is also closed under deleting bridges, Addario-Berry et al. [22], and Kang and Panagiotou [23] proved the conjecture. The full conjecture was very recently (2015) proved by Chapuy and Perarnau [24].

There is a stronger form of Theorem 3.1. Recall that X is *stochastically at most* Y, written as $X \leq_s Y$, if $\mathbb{P}(X \geq t) \leq \mathbb{P}(Y \geq t)$ for each t. If $\mathcal{G}$ is bridge-addable and $R_n \in_u \mathcal{G}$, then in fact comp(R_n) is stochastically at most $1 + \text{Po}(1)$ (see [14]). Here Po(λ) refers to the Poisson distribution with mean λ, or a corresponding random variable. This may be proved by extending the proof of Theorem 3.1; and it yields that result immediately, since

$$\mathbb{P}(R_n \text{ is connected}) = \mathbb{P}(\text{comp}(R_n) \leq 1) \geq \mathbb{P}(\text{Po}(1) \leq 0) = 1/e.$$

3.2 Bridge-addability and the fragment

The *fragment* "left over," Frag(G), is the subgraph induced on the vertices not in the biggest component of G (that is, with the maximum number of vertices). If there is more than one biggest component we break ties in some way; for example, for graphs on $[n]$ we may do so by choosing the candidate component which contains the least vertex. Let frag(G) be $v(\text{Frag}(G))$, the number of vertices in Frag(G). The following result from McDiarmid [25, 26] shows that the biggest component is typically giant!

Theorem 3.3. *If the class $\mathcal{G}$ is bridge-addable and $R_n \in_u \mathcal{G}$ then*

$$\mathbb{E}[\mathrm{frag}(R_n)] < 2.$$

To prove Theorem 3.3 we use a basic lemma on graphs from McDiarmid [25], similar to the last lemma.

Lemma 3.4. *If the graph G has n vertices, then*

$$|\mathrm{Cross}(G)| \geq (n/2) \cdot \mathrm{frag}(G).$$

Proof. An easy convexity argument shows that if $x, x_1, x_2, \ldots$ are positive integers such that each $x_i \leq x$ and $\sum_i x_i = n$ then $\sum_i \binom{x_i}{2} \leq \frac{1}{2} n(x-1)$. For, if $n = ax + y$ where $a \geq 0$ and $0 \leq y \leq x-1$ are integers, then

$$\sum_i \binom{x_i}{2} \leq a\binom{x}{2} + \binom{y}{2} \leq a\binom{x}{2} + \frac{y(x-1)}{2} = \frac{1}{2} n(x-1).$$

Hence if we denote the maximum number of vertices in a component by x, so that $\mathrm{frag}(G) = n - x$, then

$$|\mathrm{Cross}(G)| \geq \binom{n}{2} - \frac{1}{2} n(x-1) = \frac{1}{2} n(n-x) = \frac{1}{2} n \, \mathrm{frag}(G)$$

as required. $\qquad\square$

Proof of Theorem 3.3. Using Lemma 3.4 for the first inequality, and then arguing as in the proof of Theorem 3.1, we see that

$$(n/2) \sum_{G \in \mathcal{G}_n} \mathrm{frag}(G) \leq \sum_{G \in \mathcal{G}_n} |\mathrm{Cross}(G)|$$

$$\leq \sum_{G \in \mathcal{G}_n} |\mathrm{Bridge}(G)|$$

$$\leq |\mathcal{G}_n| \cdot (n-1).$$

Thus

$$\mathbb{E}[\mathrm{frag}(R_n)] \leq (2/n) \cdot (n-1) < 2$$

as required. $\qquad\square$

4 Growth constants

To go further with our investigations of $R_n \in_u \mathcal{G}$, we will need to know that the numbers $|\mathcal{G}_n|$ do not jump around too much. We say that $\mathcal{G}$ has a *growth constant* if $0 < \rho_{\mathcal{G}} < \infty$, and

$$(|\mathcal{G}_n|/n!)^{1/n} \to \rho_{\mathcal{G}}^{-1} \quad \text{as } n \to \infty,$$

that is, if
$$|\mathcal{G}_n| = (1+o(1))^n \rho_{\mathcal{G}}^{-n} \, n!.$$

It is easy to check that if $(|\mathcal{G}_n|/n!)^{1/n}$ tends to a limit then that limit must be $\rho_{\mathcal{G}}^{-1}$.

If $\mathcal{G}$ contains arbitrarily long paths, then clearly $\rho_{\mathcal{G}} \leq 1$, since $|\mathcal{G}_n|/n! \geq 1/2$ for each n. Bernardi et al. [27] showed that, if $\mathcal{G}$ is monotone (that is, closed under deleting edges or vertices) and does not contain all paths, then $\rho_{\mathcal{G}} = \infty$. Any minor-closed class is monotone.

4.1 When is there a growth constant?

The following key lemma is from McDiarmid [14, 28], and relies on supermultiplicativity.

Lemma 4.1. *Let the nonempty class $\mathcal{G}$ of graphs be addable and satisfy $\rho_{\mathcal{G}} > 0$. Then $\mathcal{G}$ has a growth constant.*

Proof. Let $\mathcal{C}$ be the class of connected graphs in $\mathcal{G}$. Let a and b be positive integers, and let $V = [a+b] = \{1, \ldots, a+b\}$. Because $\mathcal{G}$ is decomposable, we have
$$|\mathcal{G}_{a+b}| \geq \binom{a+b}{a} |\mathcal{C}_a||\mathcal{C}_b| \frac{1}{2},$$
since each graph formed by picking an a-subset A of V, and putting a graph in $\mathcal{C}$ on A and a graph in $\mathcal{C}$ on $V \setminus A$, must be in $\mathcal{G}$; and these graphs we construct are distinct except perhaps if $a = b$ (which is where the $\frac{1}{2}$ comes from). But also $\mathcal{G}$ is bridge-addable, so $|\mathcal{C}_n| \geq |\mathcal{G}_n|/e$ by Theorem 3.1; and thus
$$|\mathcal{G}_{a+b}| \geq \binom{a+b}{a} \frac{|\mathcal{G}_a|}{e} \frac{|\mathcal{G}_b|}{e} \frac{1}{2}.$$

Hence $f(n) = \frac{|\mathcal{G}_n|}{2e^2 n!}$ satisfies $f(a+b) \geq f(a) \cdot f(b)$; that is, f is *supermultiplicative*. Now we may use "Fekete's lemma" (see for example van Lint and Wilson [29]) to see that
$$f(n)^{1/n} \to \gamma := \sup_k f(k)^{1/k}$$
where γ is finite since $\rho_{\mathcal{G}} > 0$, and $\gamma > 0$ since $\mathcal{G}$ is nonempty. But as $n \to \infty$, $(|\mathcal{G}_n|/n!)^{1/n} \sim f(n)^{1/n}$, so $(|\mathcal{G}_n|/n!)^{1/n}$ must also converge to γ. $\square$

The last result, together with the result noted at the end of Section 2.1 that each proper addable minor-closed class $\mathcal{G}$ satisfies $\rho_{\mathcal{G}} > 0$, gives:

Theorem 4.2. *Each proper addable minor-closed class $\mathcal{G}$ has a growth constant.*

In particular, the class $\mathcal{P}$ of planar graphs has a growth constant. Indeed, each surface class $\mathcal{G}^S$ has a growth constant, the same as for $\mathcal{P}$,

McDiarmid [25]. This may be proved by induction on the Euler genus, using the fact that if G is embedded in a surface of positive genus then there must be a short non-contractible cycle. (We now know *much* more, as we noted at the end of Section 1.) Bernardi et al. [27] made the following natural conjecture.

Conjecture 4.3. *Each proper minor-closed class of graphs has a growth constant.*

4.2 Pendant Appearances Theorem

The subject of this section is a very useful theorem that concerns any graph class with a growth constant. Let H be a connected graph, with a specified root vertex. We say that G has a *pendant copy* of H if G contains a bridge e with H "at one end"; that is, if we remove e from G then there is a component isomorphic to H, where e was incident with the root.

It is convenient to make more precise demands. Let H be a connected graph on vertex set $[h] = \{1, \ldots, h\}$, and let G be a graph on vertex set $\{1, \ldots, n\}$, where $n > h$. Let W be an h-set of vertices of G and let the root vertex r_W be the least element in W. We say that H has a *pendant appearance* at W in G if (a) the increasing bijection from $[h]$ to W gives an isomorphism between H and the induced subgraph $G[W]$ of G; and (b) there is exactly one edge in G between W and the rest of G, and this edge is incident with the root r_W.

We call a connected rooted graph H *attachable* to $\mathcal{G}$ if whenever we have a graph G in $\mathcal{G}$ and a disjoint copy of H, and we add an edge between a vertex in G and the root of H, then the resulting graph (which has a pendant copy of H) must be in $\mathcal{G}$. For an addable minor-closed class $\mathcal{G}$, the attachable graphs are the connected rooted graphs in $\mathcal{G}$. For a surface class $\mathcal{G}^S$, the attachable graphs are the connected rooted planar graphs.

Corollary 4.5 following the next theorem is the standard version of the Pendant Appearances Theorem [14, 19]. Think of $\rho_{\mathcal{H}} > \rho_{\mathcal{G}}$ as indicating that $\mathcal{H}$ is *much* smaller than $\mathcal{G}$.

Theorem 4.4. *Let the connected graph H be attachable to the class $\mathcal{G}$, where $0 < \rho_{\mathcal{G}} < \infty$. Then there exists $\alpha > 0$ such that, if $\mathcal{H}$ is the set of graphs $G \in \mathcal{G}$ with less than αn pendant appearances of H (where $n = v(G)$), then $\rho_{\mathcal{H}} > \rho_{\mathcal{G}}$.*

Proof idea. From each $G \in \mathcal{H}_n$, by adding δn pendant appearances of H we may construct many graphs G' in $\mathcal{G}_{n'}$, where $n' = n + v(H)\delta n$. Since each G has few pendant appearances of H, each graph G' is not constructed very often; and so if $\mathcal{H}_n$ were big we would get too many graphs in $\mathcal{G}_{n'}$.

Corollary 4.5. (The Pendant Appearances Theorem) *If $\mathcal{G}$ has a growth constant, then for $R_n \in_u \mathcal{G}$*

$$\Pr(R_n \text{ has } < \alpha n \text{ pendant appearances of } H) = e^{-\Omega(n)}.$$

Proof. The probability here is

$$\frac{|\mathcal{H}_n|}{|\mathcal{G}_n|} \leq \frac{(\rho_{\mathcal{H}}^{-1} + o(1))^n}{(\rho_{\mathcal{G}}^{-1} + o(1))^n} = e^{-\Omega(n)},$$

as required.   □

4.3 Some applications of the Pendant Appearances Theorem

Vertex degrees

Suppose that the class $\mathcal{G}$ of graphs has a growth constant. If the d-vertex star (rooted at the center) is attachable to $\mathcal{G}$, then with high probability there are linear numbers of vertices of degree d. We use "with high probability" or **whp** to mean "with probability $\to 1$ as $n \to \infty$."

In particular, if the 3-vertex star (cherry) is attachable to $\mathcal{G}$, then **whp** there are at least αn pendant cherries (for some constant $\alpha > 0$), and thus has at least $2^{\alpha n}$ automorphisms. We shall use this result in Section 5 when we discuss unlabeled graphs.

Searching for a subgraph

Let $\mathcal{G}$ be an addable proper minor-closed class of graphs. Let H be a fixed connected graph in $\mathcal{G}$ with h vertices. How quickly can we find a subgraph H in R_n or verify there is no such subgraph?

An easy method takes $O(1)$ expected time. We can check in $O(1)$ time if a vertex is the root of a pendant copy of H (assuming vertices have adjacency lists). We test the vertices in turn until we succeed, or run out of vertices, in which case we run through all sets of h vertices.

The expected number of tests is at most $1/\alpha + o(1)$, so the expected time in the test phase is $O(1)$. In the second phase the expected time is $O(n^h e^{-\Omega(n)}) = o(1)$, since the probability we enter the phase is $e^{-\Omega(n)}$. Thus the overall expected time is $O(1)$. Similarly, we can seek an induced copy of H or a minor H in $O(1)$ expected time.

Coloring

For $R_n \in_u \mathcal{P}$, **whp** there is a pendant copy of K_4, thus $\omega(R_n) = 4$ and so $\chi(R_n) = 4$ (by the four-color theorem). But we can say more: there is a fast expected-time optimal coloring algorithm for random planar graphs $R_n \in_u \mathcal{P}$, as follows.

By the above remarks, we may first check if R_n has a pendant copy of K_4, in constant expected time. If there is one then we apply the quadratic time algorithm to four-color planar graphs, which follows from the proof of the four-color theorem, see Robertson et al. [30], to color the graph optimally. In the remaining cases, which happen with probability $e^{-\Omega(n)}$, we color the graph optimally in subexponential time $O(c^{\sqrt{n}})$ by using the $\sqrt{n}$-separator theorem. It follows that we can color a random planar graph optimally in quadratic expected time. This observation is due to Anusch Taraz and Michael Krivelevich, see [14]. Further, we see that we can determine $\chi(R_n)$ in constant expected time.

We do not know the full story for a general surface class $\mathcal{G}^S$. For $R_n \in_u \mathcal{G}^S$, we have $\omega(R_n) = 4$ and $\chi(R_n) \in \{4, 5\}$ **whp** since R_n has large facewidth **whp**, see Chapuy et al. [5]. Also, **whp** the list of chromatic number of R_n is 5, again see Chapuy et al. [5].

Finally here, consider Hadwiger's Conjecture, which is "probably the most famous open problem in graph theory," see Seymour [31]. It says that if $\chi(G) = k$ then G has a minor K_k. Thus the conjecture being false says that for some k, there is a graph $G_0 \in \mathrm{Ex}(K_k)$ with $\chi(G_0) \geq k$. But then G_0 is attachable to $\mathrm{Ex}(K_k)$, and by the Pendant Appearances Theorem, for $R_n \in_u \mathrm{Ex}(K_k)$ we have $\chi(R_n) \geq k$ with exponentially small failure probability: it would be hard to avoid counterexamples!

5 Unlabeled graphs

For unlabeled graphs we can follow some of the steps which worked for labeled graphs. Given a class $\mathcal{G}$ of graphs, let $\tilde{\mathcal{G}}$ denote the corresponding set of unlabeled graphs. We may think of an unlabeled n-vertex graph as an equivalence class of labeled graphs on $[n]$, under graph isomorphism. Thus the set $\mathcal{G}_n$ of graphs on $[n]$ is partitioned into the set $\tilde{\mathcal{G}}_n$ of unlabeled n-vertex graphs in $\tilde{\mathcal{G}}$.

The (ordinary) generating function for $\tilde{\mathcal{G}}$ is

$$\tilde{G}(x) = \sum_{n \geq 0} |\tilde{\mathcal{G}}_n| x^n = \sum_{G \in \tilde{\mathcal{G}}} x^{v(G)},$$

with radius of convergence $\rho_{\tilde{G}}$. We say that $\tilde{\mathcal{G}}$ has an *(unlabeled) growth constant* if $0 < \rho_{\tilde{G}} < \infty$ and $|\tilde{\mathcal{G}}_n|^{1/n} \to \rho_{\tilde{G}}^{-1}$ as $n \to \infty$. Much as in the labeled case, it is easy to check that if $|\tilde{\mathcal{G}}_n|^{1/n}$ tends to a limit, then that limit must be $\rho_{\tilde{\mathcal{G}}}^{-1}$.

The "smallness" result of Dvořák and Norine [13] mentioned at the end of Section 2.1 was stated in terms of labeled graph classes, but in fact it applies also to unlabeled graph classes. It says that, for each proper minor-closed class $\mathcal{G}$ of graphs, there is a constant c such that $|\tilde{\mathcal{G}}_n| \leq c^n$ for each n; that is, $\rho_{\tilde{\mathcal{G}}} > 0$. (This result immediately implies $\rho_{\mathcal{G}} > 0$ for the labeled case, since $|\mathcal{G}_n| \leq n! \, |\tilde{\mathcal{G}}_n|$.) The next theorem corresponds to the central result Theorem 4.2 for the labeled case.

Theorem 5.1. *Let $\mathcal{G}$ be a proper addable minor-closed class of graphs, and let C be the class of connected graphs in $\mathcal{G}$. Then the corresponding unlabeled classes $\tilde{\mathcal{G}}$ and $\tilde{\mathcal{C}}$ each have a growth constant, and $\rho_{\tilde{\mathcal{G}}} = \rho_{\tilde{\mathcal{C}}} < \rho_{\mathcal{G}}$.*

Proof. Let $\tilde{\mathcal{C}}^{\bullet}$ denote the set of (vertex-) rooted graphs in $\tilde{\mathcal{C}}$. Observe that $|\tilde{\mathcal{C}}_k| \leq |\tilde{\mathcal{C}}_k^{\bullet}| \leq k|\tilde{\mathcal{C}}_k|$, so $\limsup_{k \to \infty} |\tilde{\mathcal{C}}_k^{\bullet}|^{1/k} = \rho_{\tilde{\mathcal{C}}}^{-1} < \infty$; and thus $\sup_k |\tilde{\mathcal{C}}_k^{\bullet}|^{1/k} < \infty$.

We claim that $f(n) = |\tilde{\mathcal{C}}_n^{\bullet}|$ is supermultiplicative; that is, for positive integers a and b

$$|\tilde{\mathcal{C}}_{a+b}^{\bullet}| \geq |\tilde{\mathcal{C}}_a^{\bullet}| \cdot |\tilde{\mathcal{C}}_b^{\bullet}|.$$

To see this, note first that we may assume that $a \leq b$. We may form a graph H in $\tilde{\mathcal{C}}_{a+b}^{\bullet}$ from disjoint graphs $H_a \in \tilde{\mathcal{C}}_a^{\bullet}$ with root r_a and $H_b \in \tilde{\mathcal{C}}_b^{\bullet}$ with root r_b, by adding the edge $r_a r_b$ and making r_a the new root r^*. There is no double counting, since in the new graph H there is a unique bridge e incident with r^* such that at least half the vertices are in the component of $H \setminus e$ not containing r^*.

Since $f(n)$ is supermultiplicative, by Fekete's Lemma as before, as $n \to \infty$

$$|\tilde{\mathcal{C}}_n^{\bullet}|^{1/n} \to \gamma := \sup_k |\tilde{\mathcal{C}}_k^{\bullet}|^{1/k} < \infty.$$

We have now seen that $\tilde{\mathcal{C}}^{\bullet}$ has growth constant γ, and so this holds also for $\tilde{\mathcal{C}}$, by the inequalities $|\tilde{\mathcal{C}}_n| \leq |\tilde{\mathcal{C}}_n^{\bullet}| \leq n|\tilde{\mathcal{C}}_n|$ which we noted before. It follows that $\tilde{\mathcal{G}}$ also has growth constant γ, for example by noting that $|\tilde{\mathcal{C}}_n| \leq |\tilde{\mathcal{G}}_n| \leq |\tilde{\mathcal{C}}_{n+1}^{\bullet}|$.

It remains to show that $\rho_{\tilde{\mathcal{G}}} < \rho_{\mathcal{G}}$. Recall that the isomorphism class of a graph G in $\tilde{\mathcal{G}}_n$ has size $n!/\mathrm{aut}(G)$, where $\mathrm{aut}(G)$ is the number of automorphisms of G. We saw in the last section that most graphs in $\mathcal{G}_n$ have at least 2^{an} automorphisms, where $\alpha > 0$ is a constant. Thus most graphs in $\mathcal{G}_n$ are in isomorphism classes of size $\leq 2^{-an} n!$. Hence

$$|\tilde{\mathcal{G}}_n| \geq (1 + o(1)) \, |\mathcal{G}_n| / (2^{-an} n!),$$

that is

$$|\mathcal{G}_n| / n! \leq (1 + o(1)) \, 2^{-an} |\tilde{\mathcal{G}}_n|;$$

and it follows that $\rho_{\mathcal{G}} \geq 2^{\alpha} \rho_{\tilde{\mathcal{G}}}$. □

We could adapt the first part of the proof above to prove Theorem 4.2 on the existence of growth constants for labeled graphs. The proof we gave for that result relied on the connectivity lower bound for a bridge-addable class given in Theorem 3.1. We do not know any corresponding result for a bridge-addable class of unlabeled graphs.

For the class $\mathcal{P}$ of planar graphs, by the last theorem, $\tilde{\mathcal{P}}$ has a growth constant and (writing $\gamma_{\tilde{\mathcal{P}}}$ for $\rho_{\tilde{\mathcal{P}}}^{-1}$ and so on) $\gamma_{\tilde{\mathcal{P}}} > \gamma_{\mathcal{P}}$, as shown in McDiarmid et al. [14]. We know $\gamma_{\mathcal{P}}$ precisely – it is 27.226878 to six decimal places, and we can determine as many digits as we wish, see [3, 32]: but that is not the case for $\gamma_{\tilde{\mathcal{P}}}$. The best known bounds are $\gamma_{\mathcal{P}} < \gamma_{\tilde{\mathcal{P}}} \leq 30.061$, see McDiarmid [14, 33]. (We do not know an asymptotic formula or even smoothness for $\tilde{\mathcal{P}}$, see later.)

We noted that, for each given surface S, the class $\mathcal{G}^S$ of graphs embeddable in S has a growth constant, namely the planar growth constant $\gamma_{\mathcal{P}} = \rho_{\mathcal{P}}^{-1}$. Similarly, for each fixed surface S, the class $\tilde{\mathcal{G}}^S$ of unlabeled graphs embeddable on S has growth constant $\gamma_{\tilde{\mathcal{P}}}$, see McDiarmid [25]. There is a natural conjecture corresponding to Conjecture 4.3 for the labeled case.

Conjecture 5.2. *For each proper minor-closed class $\mathcal{G}$, the corresponding class $\tilde{\mathcal{G}}$ of unlabeled graphs has a growth constant, and indeed the class $\tilde{\mathcal{C}}$ of connected graphs in $\tilde{\mathcal{G}}$ has a growth constant.*

(The second part of the conjecture would imply the first part, using standard results on generating functions of multisets.)

6 Smoothness

Next we take a step beyond just asking for $\mathcal{G}$ to have a growth constant. Graph classes such as the planar graphs $\mathcal{P}$ are "smooth": we shall see that for such classes we can say more, for example about the probability of being connected.

Let $\mathcal{G}$ be any class of graphs with $0 < \rho_{\mathcal{G}} < \infty$. Suppose that $\mathcal{G}_n \neq \emptyset$ for n sufficiently large, say for $n \geq m$. Let $r_n = n|\mathcal{G}_{n-1}|/|\mathcal{G}_n|$. If we assume that isolated vertices can be freely added to graphs in $\mathcal{G}$, then r_n is the expected number of isolated vertices in R_n, where $R_n \in_u \mathcal{G}$. Observe that for $n > m$

$$\frac{|\mathcal{G}_n|}{n!} = \frac{|\mathcal{G}_m|}{m!} \prod_{k=m+1}^{n} r_k^{-1}$$

Using this equation it is easy to see that

$$\liminf_{n \to \infty} r_n \leq \rho_{\mathcal{G}} \leq \limsup_{n \to \infty} r_n. \tag{6.1}$$

We call $\mathcal{G}$ *smooth* if r_n tends to a limit as $n \to \infty$. In this case the limit clearly must be $\rho_{\mathcal{G}}$ and $\mathcal{G}$ must have a growth constant.

Classes such as forests, series-parallel graphs, planar graphs $\mathcal{P}$ and the surface class $\mathcal{G}^S$, for which we know an asymptotic counting formula are all smooth. Showing smoothness is often a crucial step in proving results about $R_n \in_u \mathcal{G}$.

6.1 Smoothness for surface and addable minor-closed classes

We noted earlier that each surface class $\mathcal{G}^S$ has growth constant $\gamma_\mathcal{P}$, the planar graph growth constant. Bender et al. [34] showed that $\mathcal{G}^S$ is smooth. The proof did not involve an asymptotic counting formula (and indeed none was then known). Their composition method will tell us more.

The key idea in the proof involves the core. The *core* (or 2-core) of a graph G is the unique maximal subgraph with minimum degree of at least 2: it may be obtained from G by repeatedly stripping off leaves. Let $\mathcal{G}^{\delta \geq 2}$ denote the class of graphs G in $\mathcal{G}$ with minimum degree $\delta(G) \geq 2$. The idea is that if $\mathcal{G}^{\delta \geq 2}$ grows reasonably smoothly then rooting trees in the core leads to a smooth class $\mathcal{G}$. Further, when we show in this way that $\mathcal{G}$ is smooth, we also learn about the typical size of the core.

Theorem 6.1. *([28, 34]) Let $\mathcal{G}$ either be the class $\mathcal{G}^S$ of graphs embeddable in a given surface S, in which case set $\mathcal{A} = \mathcal{P}$; or a proper addable minor-closed class of graphs, in which case set $\mathcal{A} = \mathcal{G}$. Let $\mathcal{C}$ be the class of connected graphs in $\mathcal{G}$.*

Then $\mathcal{G}$ and $\mathcal{C}$ are smooth. Further, let ρ_2 be the radius of convergence of $\mathcal{A}^{\delta \geq 2}$. Let $R_n \in_u \mathcal{G}$ or $R_n \in_u \mathcal{C}$: then for any $\varepsilon > 0$

$$\mathbb{P}(|v(\text{core}(R_n)) - (1 - \rho_2)n| > \varepsilon n) = e^{-\Omega(n)}.$$

Perhaps we have smoothness more generally?

Conjecture 6.2. *Every proper minor-closed class is smooth.*

In the next subsections, we will see what smoothness will yield.

6.2 Boltzmann Poisson random graph

Let the class $\mathcal{G}$ be decomposable. Fix $\rho > 0$ such that $G(\rho)$ is finite; and let

$$\mu(H) = \frac{\rho^{v(H)}}{\text{aut}(H)} \quad \text{for each } H \in \tilde{\mathcal{G}}. \tag{6.2}$$

(Recall that $\tilde{\mathcal{G}}$ denotes the set of unlabeled graphs in $\mathcal{G}$.) We will normalize the quantities μ_H to give probabilities. Since each graph $H \in \tilde{\mathcal{G}}_n$ consists of $\frac{n!}{\text{aut}(H)}$

isomorphic graphs $G \in \mathcal{G}_n$, we have

$$\frac{x^n}{\mathrm{aut}(H)} = \sum_{G \in H} \frac{\mathrm{aut}(H)}{n!} \frac{x^n}{\mathrm{aut}(H)} = \sum_{G \in H} \frac{x^n}{n!}.$$

Therefore

$$G(x) = \sum_{G \in \mathcal{G}} \frac{x^{v(G)}}{v(G)!} = \sum_{H \in \tilde{\mathcal{G}}} \sum_{G \in H} \frac{x^{v(G)}}{v(G)!} = \sum_{H \in \tilde{\mathcal{G}}} \frac{x^{v(H)}}{\mathrm{aut}(H)}.$$

Thus

$$G(\rho) = \sum_{H \in \tilde{\mathcal{G}}} \mu(H). \tag{6.3}$$

The *Boltzmann Poisson random graph* $R = BP(\mathcal{G}, \rho)$ takes values in $\tilde{\mathcal{G}}$, with

$$\mathbb{P}(R = H) = \frac{\mu(H)}{G(\rho)} \quad \text{for each } H \in \tilde{\mathcal{G}}.$$

Let $\mathcal{C}$ denote the class of connected graphs in $\mathcal{G}$. Then $G(x) = e^{C(x)}$ by the *exponential formula*, see for example [2]. For each $H \in \tilde{\mathcal{C}}$, the unlabeled graphs in $\mathcal{C}$, let $\mathrm{comp}(G, H)$ denote the number of components of G isomorphic to H.

Proposition 6.3. *The random variables* $\mathrm{comp}(R, H)$ *for* $H \in \tilde{\mathcal{C}}$ *are independent, with* $\mathrm{comp}(R, H) \sim \mathrm{Po}(\mu(H))$.

Proof. Each sum and product below is over all H in $\tilde{\mathcal{C}}$. Let the unlabeled graph G consist of n_H components isomorphic to H for each $H \in \tilde{\mathcal{C}}$, where $0 \le \sum_H n_H < \infty$. Then

$$\rho^{v(G)} = \prod_H \rho^{v(H)n_H}$$

and

$$\mathrm{aut}(G) = \prod_H \mathrm{aut}(H)^{n_H} n_H!.$$

Also since $\sum_H \mu(H) = C(\rho)$ by (6.3) applied to $\mathcal{C}$,

$$\frac{1}{G(\rho)} = e^{-C(\rho)} = \prod_H e^{-\mu(H)}.$$

Hence, for each $G \in \tilde{\mathcal{G}}$,

$$\mathbb{P}(R = G) = e^{-C(\rho)} \frac{\rho^{v(G)}}{\mathrm{aut}(G)}$$

$$= \prod_H e^{-\mu(H)} \frac{\mu(H)^{n_H}}{n_H!}$$

$$= \prod_H \mathbb{P}(\mathrm{Po}(\mu(H)) = n_H).$$

Thus the probability factors appropriately, and the random variables $\text{comp}(R, H)$ for $H \in \tilde{C}$ satisfy

$$\mathbb{P}(\text{comp}(R, H) = n_H \ \forall H \in \tilde{C}) = \prod_H \mathbb{P}(\text{comp}(R, H) = n_H).$$

This holds for every choice of non-negative integers n_H with $\sum_{H \in \tilde{C}} n_H < \infty$, and thus also without this last restriction (since both sides are zero if the sum is infinite). This completes the proof. $\qquad\qquad\qquad\qquad\qquad\qquad\square$

6.3 The fragment for surface and addable minor-closed classes

Recall that the fragment $\text{Frag}(G)$ of a graph G is the graph remaining when we discard the largest component (breaking ties some way). In particular, $\text{Frag}(G)$ is empty if and only if G is connected. Recall also that for a decomposable graph class $\mathcal{A}$, the Boltzmann Poisson random graph $BP(\mathcal{A}, \rho)$ is well defined as long as $0 < A(\rho) < \infty$.

Here is the Fragments Theorem [28] for a surface class or an addable minor-closed class: it is the reason we are interested in $BP(\mathcal{A}, \rho)$.

Theorem 6.4. *Either let $\mathcal{G}$ be $\mathcal{G}^S$ for a given surface S, in which case set $\mathcal{A} = \mathcal{P}$; or let $\mathcal{G}$ be a proper addable minor-closed class of graphs, in which case set $\mathcal{A} = \mathcal{G}$. Then $0 < A(\rho_{\mathcal{A}}) < \infty$, and the Boltzmann Poisson random graph $BP(\mathcal{A}, \rho_{\mathcal{A}})$ is well defined; and for $R_n \in_u \mathcal{G}$, the random unlabeled fragment converges in distribution to $BP(\mathcal{A}, \rho_{\mathcal{A}})$.*

Corollary 6.5. *Given distinct unlabeled connected graphs $H_1, \ldots, H_k$ in $\mathcal{A}$, the k random variables $\text{comp}(\text{Frag}(R_n), H_i)$ are asymptotically independent with distribution $\text{Po}(\mu(H_i))$, for $i = 1, \ldots, k$. In particular*

$$\mathbb{P}(R_n \text{ is connected}) \to e^{-C(\rho_{\mathcal{A}})} \quad as \ n \to \infty,$$

where C is the class of connected graphs in $\mathcal{A}$.

Consider the classes $\mathcal{T}$ of trees and $\mathcal{F}$ of forests, each with radius of convergence e^{-1}. For $R_n \in_u \mathcal{F}$, since $T(e^{-1}) = \frac{1}{2}$,

$$\mathbb{P}(R_n \text{ is connected }) = \frac{|\mathcal{T}_n|}{|\mathcal{F}_n|} \to e^{-\frac{1}{2}} \quad as \ n \to \infty,$$

as we saw in Section 3.1.

Now consider a simple example which provides a contrast to Theorem 6.4 and Corollary 6.5. Let $\mathcal{C}$ be the class of paths, and let $\mathcal{G}$ be the class of *path forests* (forests in which every component is a path). Clearly $|\mathcal{C}_n| = \frac{1}{2}n!$ for $n \geq 2$, so $\mathcal{C}$ has growth constant 1, and thus also $\mathcal{G}$ has growth constant 1 by the exponential formula. Observe that $\mathcal{G}$ is minor-closed and decomposable but not bridge-addable, and so Theorem 6.4 does not apply: indeed the conclusions of that theorem do not hold, and in particular $G(\rho_{\mathcal{G}})$ is clearly infinite. Further, for $R_n \in_u \mathcal{G}$ we have, see McDiarmid [28], that

$$\sqrt{2/n}\,\mathrm{comp}(R_n) \to 1 \quad \text{in probability as } n \to \infty. \tag{6.4}$$

Similar results hold for example for caterpillar forests, which have growth constant $\gamma \approx 1.76$, see Bernardi et al. [27]. See Bousquet-M'elou and Weller [35] for many further related results.

6.4 Smoothness and unlabeled graphs

Call a set $\tilde{\mathcal{G}}$ of unlabeled graphs *smooth* if $0 < \rho_{\tilde{\mathcal{G}}} < \infty$ and $r_n = |\tilde{\mathcal{G}}_{n-1}|/|\tilde{\mathcal{G}}_n|$ tends to $\rho_{\tilde{\mathcal{G}}}$ as $n \to \infty$. It is easy to check that if r_n tends to a limit as $n \to \infty$ then that limit must be $\rho_{\tilde{\mathcal{G}}}$, and $\tilde{\mathcal{G}}$ has a growth constant.

We know asymptotic counting formulae for a few unlabeled graph classes, for example the outerplanar graphs and the series-parallel graphs, see Bodirsky et al. [36], and these classes are smooth. But we have no general results that certain sorts of unlabeled graph classes are smooth, and in particular no result corresponding to Theorem 6.1. In the following conjecture, it would be a big step forward to establish smoothness even, for example, in the addable case.

Conjecture 6.6. *Every proper minor-closed class of unlabeled graphs is smooth.*

However, if we *assume* smoothness, then we find parallels to results on the fragment for labeled graphs, see Weller [37]. Let $\tilde{\mathcal{G}}$ be an unlabeled decomposable graph class, and suppose that $\tilde{\mathcal{G}}$ is smooth. Let $\tilde{\mathcal{C}}$ be the corresponding class of connected graphs. Let $H_1, \ldots, H_j$ be distinct graphs in $\tilde{\mathcal{C}}$, and let $k_1, \ldots, k_j$ be non-negative integers. Then the number of graphs in $\tilde{\mathcal{G}}_n$ with at least k_i components H_i for $i = 1, \ldots, j$ equals the number of graphs in $\tilde{\mathcal{G}}_{n'}$, where $n' = n - \sum_i k_i v(H_i)$; and since we are assuming that $\tilde{\mathcal{G}}$ is smooth,

$$|\tilde{\mathcal{G}}_{n'}| \sim |\tilde{\mathcal{G}}_n|\,\rho_{\tilde{\mathcal{G}}}^{\sum_{i=1}^{j} k_i v(H_i)} = |\tilde{\mathcal{G}}_n| \prod_{i=1}^{j} (\rho_{\tilde{\mathcal{G}}}^{v(H_i)})^{k_i}.$$

Now let $\tilde{R}_n \in_u \tilde{\mathcal{G}}$; that is, $\tilde{R}_n$ is uniformly distributed over the n-vertex graphs in $\tilde{\mathcal{G}}$. Let $X_i = \mathrm{comp}(\tilde{R}_n, H_i)$, the number of components of $\tilde{R}_n$ isomorphic to H_i. Then

$$\mathbb{P}(X_i \geq k_i \text{ for each } i = 1, \ldots, j) = \frac{|\tilde{\mathcal{G}}_{n'}|}{|\tilde{\mathcal{G}}_n|} \sim \prod_{i=1}^{j} (\rho_{\tilde{\mathcal{G}}}^{v(H_i)})^{k_i}.$$

It follows that, as $n \to \infty$ $X_1, \ldots, X_j$ are asymptotically independent; and X_i is asymptotically geometrically distributed with parameter $1 - \rho_{\tilde{\mathcal{G}}}^{v(H_i)}$, that is, for each fixed $t = 0, 1, \ldots$

$$\mathbb{P}(X_i = t) \to (\rho_{\mathcal{G}}^{v(H_i)})^t (1 - \rho_{\mathcal{G}}^{v(H_i)}) \quad \text{as } n \to \infty.$$

Given $\rho > 0$ such that $\tilde{G}(\rho)$ is finite, we may define the *Boltzmann geometric random graph* $\tilde{R} = BG(\tilde{G}, \rho)$ by setting

$$\mathbb{P}(\tilde{R} = H) = \frac{\rho^{v(H)}}{\tilde{G}(\rho)} \quad \text{for } H \in \tilde{G}.$$

It is straightforward to check that the random variables $\mathrm{comp}(\tilde{R}, H)$ for $H \in \tilde{C}$ are independent and $\mathrm{comp}(\tilde{R}, H)$ has the geometric distribution with parameter $1 - \rho^{v(H)}$. It is natural to conjecture that, for suitable classes $\tilde{G}$, the fragment of $\tilde{R}_n$ should converge in distribution to $BG(\tilde{G}, \rho_{\tilde{G}})$.

7 Concluding remarks

We have considered random graphs from a minor-closed class, focusing on the addable case (when the excluded minors are 2-connected), and on surface classes G^S, which are "close" to the addable class P of planar graphs. Much of what we have done may be extended to graphs sampled from weighted classes, not necessarily uniformly, see [26, 38]. Also, we can make progress in certain cases with disconnected excluded minors, for example with the class $\mathrm{Ex}((k + 1)C_3)$, which consists of the graphs with at most k vertex-disjoint cycles, see [38–42].

Acknowledgments I have been much helped by discussions of this material over recent years with many colleagues, in particular Louigi Addario-Berry, Stefanie Gerke, Mihyun Kang, Valentas Kurauskas, Marc Noy, Kostas Panagiotou, Bruce Reed, Angelika Steger, Kerstin Weller, Dominic Welsh (and apologies for omissions!).

References

[1] A. Denise, M. Vasconcellos, and D. Welsh, The random planar graph, *Congr. Numer.* **113** (1996), 61–79.

[2] P. Flajolet and R. Sedgewick, *Analytic Combinatorics*, Cambridge University Press, Cambridge, 2009.

[3] O. Giménez and M. Noy, Asymptotic enumeration and limit laws of planar graphs, *J. Amer. Math. Soc.*, **22** (2009), 309–329.

[4] E.A. Bender, Z. Gao, and N.C. Wormald, The number of labeled 2-connected planar graphs, *Electron. J. Combin.*, **9** (2002), #R43.

[5] G. Chapuy, E. Fusy, O. Giménez, B. Mohar, and M. Noy, Asymptotic enumeration and limit laws for graphs of fixed genus, *J. Combin. Theory A*, **118** (2011), 748–777.

[6] E. Bender and Z. Gao, Asymptotic enumeration of labelled graphs with a given genus, *Electron. J. Combin.*, **18** (2011), #P13.

[7] N. Robertson and P.D. Seymour, Graph minors I–XX, *J. Combin. Theory B* (1983–2004).

[8] R. Diestel, *Graph Theory*, 4th edn., Springer-Verlag, Heidelberg, 2010.

[9] W. Mader, Homomorphiesätze für Graphen, *Math. Ann.*, **178** (1968), 154–168.

[10] A.V. Kostochka, The minimum Hadwiger number for graphs of a given mean degree of vertices (in Russian), *Metody Diskret. Anal.*, **38** (1982), 37–58.

[11] A. Thomason, An extremal function for contractions of graphs, *Math. Proc. Cambridge Philos. Soc.*, **95** (1984), 261–265.

[12] S. Norine, P. Seymour, R. Thomas, and P. Wollan, Proper minor-closed families are small, *J. Combin. Theory B*, **96** (2006), 754–757.

[13] Z. Dvořák and S. Norine, Small graph classes and bounded expansion, *J. Combin. Theory B*, **100** (2010), 171–175.

[14] C. McDiarmid, A. Steger, and D. Welsh, Random planar graphs, *J. Combin. Theory B*, **93** (2005), 187–206.

[15] A. Rényi, Some remarks on the theory of trees, *Publications of the Mathematical Institute of the Hungarian Academy of Sciences*, **4** (1959), 73–85.

[16] M. Drmota, *Random Trees*, Springer, 2009.

[17] J.W. Moon, *Counting Labelled Trees*, Canadian Mathematical Monographs **1** (1970).

[18] A. Rényi, On the enumeration of trees, *Combinatorial Structures and Their Applications*, R. Guy, H. Hanani, N. Sauer, and J. Schonheim (Eds), Gordon and Breach, New York, 1970, 355–360.

[19] C. McDiarmid, A. Steger, and D. Welsh, Random graphs from planar and other addable classes, *Topics in Discrete Mathematics*, M. Klazar, J. Kratochvil, M. Loebl, J. Matousek, R. Thomas, and P. Valtr (Eds), Algorithms and Combinatorics 26. Springer, 2006, 231–246.

[20] P. Balister, B. Bollobás, and S. Gerke, Connectivity of addable graph classes, *J. Combin. Theory B*, **98** (2008), 577–584.

[21] P. Balister, B. Bollobás, and S. Gerke, Connectivity of random addable graphs, *Proc. ICDM 2008*, **13** (2010), 127–134.

[22] L. Addario-Berry, C. McDiarmid, and B. Reed, Connectivity for bridge-addable monotone graph classes, *Combin. Prob. Comput.*, **21** (2012), 803–815.

[23] M. Kang and K. Panagiotou, On the connectivity of random graphs from addable classes, *J. Combin. Theory B*, **103** (2013), 306–312.

[24] G. Chapuy and G. Perarnau, Connectivity in bridge-addable graph classes: the McDiarmid-Steger-Welsh conjecture, arXiv:1238952 [math.CO] April 2015.

[25] C. McDiarmid, Random graphs on surfaces, *J. Combin. Theory B*, **98** (2008), 778–797.

[26] C. McDiarmid, Connectivity for random graphs from a weighted bridge-addable class, *Electronic J. Combin.*, **19**(4) (2012), P53.

[27] O. Bernardi, M. Noy, and D. Welsh, Growth constants of minor-closed classes of graphs, *J. Combin. Theory B*, **100** (2010), 468–484.

[28] C. McDiarmid, Random graphs from a minor-closed class, *Combin. Prob. Comput.*, **18** (2009), 583–599.

[29] J.H. van Lint and R.M. Wilson, *A Course in Combinatorics*, 2nd edn., Cambridge University Press, Cambridge, 2001.

[30] N. Robertson, D. Sanders, P.D. Seymour, and R. Thomas, The four-color theorem, *J. Combin. Theory B*, **70** (1997), 2–44.

[31] P.D. Seymour, Hadwiger's Conjecture, manuscript, 2015.

[32] O. Giménez and M. Noy, Counting planar graphs and related families of graphs, in *Surveys in Combinatorics 2009*, 169–329, Cambridge University Press, Cambridge, 2009.

[33] N. Bonichon, C. Gavoille, N. Hanusse, D. Poulalhon, and G. Schaeffer, Planar graphs, via well-orderly maps and trees, *Graphs Combin.*, **22** (2) (2006), 185–202.

[34] E.A. Bender, E.R. Canfield, and L.B. Richmond, Coefficients of functional compositions often grow smoothly, *Electron. J. Combin.*, **15** (2008), #R21.

[35] M. Bousquet-Mélou and K. Weller, Asymptotic properties of some minor-closed classes of graphs, *Combin. Prob. Comput.*, **23** (5) (2014), 749–795.

[36] M. Bodirsky, O. Giménez, M. Kang, and M. Noy, Enumeration and limit laws for series-parallel graphs, *European Journal of Combinatorics*, **28** (2007), 2091–2105.

[37] K. Weller, Connectivity and related properties for graph classes, DPhil thesis, Oxford University, 2013.

[38] C. McDiarmid, Random graphs from a weighted minor-closed class, *Electronic J. Combin.*, **20** (2) (2013), P52, 39 pages.

[39] M. Kang and C. McDiarmid, Random unlabelled graphs containing few disjoint cycles, *Random Structures Algorithms*, **38** (2011), 174–204.

[40] V. Kurauskas and C. McDiarmid, Random graphs with few disjoint cycles, *Combin. Prob. Comput.*, **20** (2011), 763–775.

[41] V. Kurauskas and C. McDiarmid, Random graphs containing few disjoint excluded minors, *Random Structures Algorithms*, **44** (2) (2014), 240–268.

[42] C. McDiarmid, On graphs with few disjoint *t*-star minors, *European J. Combin.*, **32** (2011), 1394–1406.

[43] V. Kurauskas, On graphs containing few disjoint excluded minors, asymptotic number and structure of graphs containing few disjoint minors K4, arXiv: 1504.08107v1 [math.CO] 2015.

Index

Printed in the United States
by Baker & Taylor Publisher Services